Assessing and Measuring Statistics Cognition in Higher Education Online Environments:

Emerging Research and Opportunities

Justin P. Chase
University at Albany (SUNY), USA

Zheng Yan
University at Albany (SUNY), USA

A volume in the Advances in Higher Education and Professional Development (AHEPD) Book Series

IGI Global
DISSEMINATOR OF KNOWLEDGE

www.igi-global.com

Published in the United States of America by
IGI Global
Information Science Reference (an imprint of IGI Global)
701 E. Chocolate Avenue
Hershey PA, USA 17033
Tel: 717-533-8845
Fax: 717-533-8661
E-mail: cust@igi-global.com
Web site: http://www.igi-global.com

Library of Congress Cataloging-in-Publication Data

Names: Chase, Justin P., 1987- author. | Yan, Zheng, 1972- author.
Title: Assessing and measuring statistics cognition in higher education online environments : emerging research and opportunities / by Justin P. Chase and Zheng Yan.
Description: Hershey, PA : Information Science Reference, [2017]
Identifiers: LCCN 2017001538| ISBN 9781522524205 (hardcover) | ISBN 9781522524212 (ebook)
Subjects: LCSH: Education, Higher--Computer-assisted instruction. | Statistics--Study and teaching (Higher) | Statistics--Study and teaching--Psychological aspects. | Computer-assisted instruction--Psychological aspects. | Statistical literacy. | Cognitive learning. | Educational psychology.
Classification: LCC QA276.18 .C465 2017 | DDC 519.5071/1--dc23 LC record available at https://lccn.loc.gov/2017001538

This book is published in the IGI Global book series Advances in Higher Education and Professional Development (AHEPD) (ISSN: 2327-6983; eISSN: 2327-6991)

British Cataloguing in Publication Data
A Cataloguing in Publication record for this book is available from the British Library.

Advances in Higher Education and Professional Development (AHEPD) Book Series

ISSN:2327-6983
EISSN:2327-6991

Editor-in-Chief: Jared Keengwe, University of North Dakota, USA

MISSION

As world economies continue to shift and change in response to global financial situations, job markets have begun to demand a more highly-skilled workforce. In many industries a college degree is the minimum requirement and further educational development is expected to advance. With these current trends in mind, the **Advances in Higher Education & Professional Development (AHEPD) Book Series** provides an outlet for researchers and academics to publish their research in these areas and to distribute these works to practitioners and other researchers.

AHEPD encompasses all research dealing with higher education pedagogy, development, and curriculum design, as well as all areas of professional development, regardless of focus.

COVERAGE

- Adult Education
- Assessment in Higher Education
- Career Training
- Coaching and Mentoring
- Continuing Professional Development
- Governance in Higher Education
- Higher Education Policy
- Pedagogy of Teaching Higher Education
- Vocational Education

IGI Global is currently accepting manuscripts for publication within this series. To submit a proposal for a volume in this series, please contact our Acquisition Editors at Acquisitions@igi-global.com or visit: http://www.igi-global.com/publish/.

The Advances in Higher Education and Professional Development (AHEPD) Book Series (ISSN 2327-6983) is published by IGI Global, 701 E. Chocolate Avenue, Hershey, PA 17033-1240, USA, www.igi-global.com. This series is composed of titles available for purchase individually; each title is edited to be contextually exclusive from any other title within the series. For pricing and ordering information please visit http://www.igi-global.com/book-series/advances-higher-education-professional-development/73681. Postmaster: Send all address changes to above address.

Titles in this Series

For a list of additional titles in this series, please visit:
http://www.igi-global.com/book-series/advances-higher-education-professional-development/73681

Student Culture and Identity in Higher Education
Ambreen Shahriar (University of Sindh, Pakistan) and Ghazal Kazim Syed (University of Sindh, Pakistan)
Information Science Reference • © 2017 • 366pp • H/C (ISBN: 9781522525516) • US $185.00

Examining the Potential for Response to Intervention (RTI) Delivery Models in Secondary...
Pam Epler (Grand Canyon University, USA)
Information Science Reference • © 2017 • 111pp • H/C (ISBN: 9781522521877) • US $115.00

Fostering Sustained Learning Among Undergraduate Students Emerging Research and...
Caroline Chemosit (University of Kabianga, Kenya) John Rugutt (Illinois State University, USA) and Joseph K. Rugutt (Missouri State University–West Plains, USA)
Information Science Reference • © 2017 • 185pp • H/C (ISBN: 9781522522713) • US $125.00

Enhancing Academic Research With Knowledge Management Principles
Dhananjay S. Deshpande (Symbiosis International University, India) Narayan Bhosale (Central University of Karnataka, India) and Rajesh Jagannathrao Londhe (Camp Education Society's Rasikla M. Dhariwal Institute of Management, India)
Information Science Reference • © 2017 • 289pp • H/C (ISBN: 9781522524892) • US $200.00

Student Activism as a Vehicle for Change on College Campuses Emerging Research and...
Michael T. Miller (University of Arkansas, USA) and David V. Tolliver III (University of Arkansas, USA)
Information Science Reference • © 2017 • 118pp • H/C (ISBN: 9781522521730) • US $120.00

For an enitre list of titles in this series, please visit:
http://www.igi-global.com/book-series/advances-higher-education-professional-development/73681

IGI GLOBAL
DISSEMINATOR OF KNOWLEDGE

www.igi-global.com

701 East Chocolate Avenue, Hershey, PA 17033, USA
Tel: 717-533-8845 x100 • Fax: 717-533-8661
E-Mail: cust@igi-global.com • www.igi-global.com

Table of Contents

Preface

Students are generally unaware of the range of variation in thinking about statistics, learning statistics and using statistics professionally, and tend to assume that their fellow students think in the same way that they do. - Petocz & Reid, 2005, p. 798

I earned a "*D* +" in my first University level statistics course, just short of the "*C* -" I needed to move on with my course work. For the first time in my life, I was required to retake a class. Upon the second completion I surpassed this threshold, earning an "*A* -", but I still had no love for the discipline. Statistics was dry, uninteresting, and - worst of all - generally taught early in the morning. At the time, I saw it as an obstacle, standing in the way of earning my degree. Instead of attempting to absorb information that could become a lifetime skill, I relied on my source memory to encode pages of statistical formula. I did not really care what these formula meant because I could not foresee how statistics could be of much use to me, or anyone else for that matter.

The moment I began to love statistics was upon completion of a my third graduate level statistics course. Initially I have reservations about this course as it was my first statistics course in sometime, and it was taught 8:45AM. However, as I attended classes - ready with copious amounts of coffee - became fascinated by the systematic framework that statistics allows one to fold onto a seemingly chaotic world. This was not only the first time I felt I understood statistics, but the first time I developed a deep appreciation for how the educator was able to relay highly complex information using simple examples, and assessed our knowledge using a highly aligned series of assessments.

I became so excited at having learned these new skills, I applied them every way I could. I computed an ANOVA to determine the type of food my cat liked to eat, and attempted to develop predictive models to determine how weather influenced when the City bus would arrive. As I reflected on these new skills I felt that the educator had made the difference. If not for her clarity, I would not have discovered these latent skills, nor would I have felt competent in their application. I began to dissect this individual's teaching method, and strongly felt that I have witnessed the *best* method for learning and teaching statistics.

However, this perception would change drastically in the following term, when I was assigned to be the teaching assistant for that same statistics class. At first I was excited at the thought of working with this professor whose teaching method I grew to admire. However, this educator was in the process of retiring, and a professor whom I had never met was assigned to take their place. While I had heard marvelous things about the educator taking over the class, I simply assumed that they would attempt to replicate the class I had taken. I felt that the previous professor had perfect the method of teaching statistics, and so why would anyone make changes? However, upon receiving the syllabus for this new course, I was dumbstruck.

While the syllabus I received was for the statistics class I had taken, it was clearly not the same course. Students would hand in a weekly assignment that they required them to identify scholarly articles that used the statistical analyses the class was learning about that week. Further, students also had to complete a computer activity involving the statistical software SPSS. There were no exams and the course demanded that students reflect and make use of self-assessments. Where were the worksheets where students applied formula?

Upon attending the first class, my shocked demeanor was prolonged. This was not a face-to-face statistics class. Void of much direct lecturing, class time was often used to have students practice the skills they would need to complete that weeks homework. Whether it was finding an article, or using SPSS, I walked around the room with the professor answering questions. As I think back on this course, I feel it can only be described as chaotic relative to the traditional systematic approach I had observed. However, I soon began to learn that this chaos was not only intended, but educationally beneficial.

As the term progressed, I directly observed students becoming faster, and more effective in finding articles, computing analyses with SPSS, and applying statistical practices with greater accuracy. The quality of content reflected on student homework increased, and I became amazed with the seemingly increasing level of complexity their questions denoted. I now understood why the class was structured the way it was. This professor was teaching a very specific type of statistics course, geared toward the abilities and needs this highly diverse class of graduate students. But beyond simply understanding why this method was used, I found myself developing a great appreciation and becoming a huge advocate for this method.

Again, excited by this new format of statistics, I presumed all students must love this class. While it lacked traditional formatting and was arguably more skill - than content - oriented, it seemed likely to serve students well for their lives. Yet, I reflected how in both courses, a small set of students conveyed their displeasure. In that first course, students argued the educator was too demanding and too technical. In the second course, some students felt that the material did not incorporate formulas and relied too much on SPSS. In both instances, a very small proportion (less than 1%) seemed to remain unhappy.

This was the moment I began to understand statistics education. Statistics is not simply what someone tells another, but rather, how each individual sees and understands empirical information. For so long, I had sought to find the best method of teaching statistics, that I forgot that there might not be a *best* method. These two experiences created dissonance in my view of statistics.

I felt that these educators were both magnificent, but could also appreciate where students were coming from, for I remember a time I felt that I could only learn statistics under the conditions that were right to me. However, as I saw these two educators in action, their methods were secondary, as from an educational standpoint, they both did a lot of things exceedingly well. These things, which will be highlighted by the educational psychological findings reviewed throughout this book, were fairly simple. Both educators seemed to have a fundamental understanding of learning statistics efficiently and effectively has a lot to do with how students think about statistics (Chapter 2), how statistical cognitive processes influence student learning (Chapter 3), what type of environments are conducive to developing statistics cognition (Chapter 4), when emotions, moods, and attitudes can enhance or harm statistics cognition (Chapter 5), and as with that small percentage in each statistics course who reported dislike for the methods, how meaningful individual differences modulate statistics cognitions (Chapter 6).

Ultimately, what I learned from these experiences was that most students could learn a lot about statistics regardless of the specific method used, provided the educator followed key educational principles such as using formative assessments, and providing adequate correctives so students could capitalize on errors they made. While there may be some students who have strong preferences for one method, or another, this experience helped me understand that statistical understanding was not innate, regardless of how easy others make it look, or claim it should be. To fully understand statistics, we must be experienced with the appropriate domain specific skill sets, so that we may be able to make creative use out of the empirically supported methods that the field is founded upon.

Although this lesson about the true meaning of statistics has proven to be invaluable to me, I must also acknowledge the support of Dr. Yan during this process. None of this book would have been possible had it not been for his willingness to provide the opportunity to work on an independent project, that resulted in large scale review of the field of statistics education. After spending so much time in academia, it is easy to become burnt out, and lose focus on the drive that originally lead down this path, and I thank him for helping me rediscover this interest.

In addition to Dr. Yan's support as a scholarly mentor, I was also inspired by his pursuit of new and exciting research endeavors. Already Dr. Yan has become a well established expert in the fields of cyber behavior and mobile psychology, and I strove to mimic his keen ability to identify emerging literatures while synthesizing their theoretical impacts. In essence, it was not until I worked with Dr. Yan that I realized that when undertaking a psychological project one must be ever cognizant of the future, and focus our efforts into identifying what we believe will be of great value to the fields that interest us. Often, in graduate school, it is easy to lose sight of the contribution that we are expected to make as scientists, and become complacent in the comfort of the empirical familiar.

In addition to the lessons learned from my observations and work with Dr. Yan, the work presented within this book would not have been possible without the editors and associates at IGI Global; Courtney Tychinski, and Jan Travers. I was somewhat hesitant to undertake this project at first, but upon seeing the call for proposals for the *Emerging Research and Opportunities* series, I felt the need to seize this opportunity. While working with IGI, I certainly felt that they were just as excited about this book as I was, and it was refreshing to work with such a dedicated team dedicated to providing quality feedback. Without the support and thoughtful comments of the editors at IGI, this book would not have been as concise, poignant, or as meaningful as is the final product that you are now reading.

Further, this work could not have been possible without the dedication of great scholars. In particular, the collective works of Dr. Ann L. Brown have highlighted the necessity of design based research, and provided a systematic outline for how evidence can be collected within a classroom. This, coupled with Dr. J. Dewey's century old theoretical arguments about the nature of schooling (see Chapter 1), have tremendously contributed to my understanding and philosophy of both education, and science. Additionally, Dr. Joan Garfield's theoretical work on learning statistics is, if nothing else, directly responsible for the culmination of modern statistics education. Further, the works of Daniel Kahneman and Amos Tversky have been of particular inspiration for not only showing how far reaching the magnitude of research on statistics cognition can be, but for their dedication to designing exceedingly simple - and extraordinarily clever - experiments. Finally, the writings of Nate Silver have also served as an extraordinary inspiration, as it was my first glimpse at understanding the practical world of dynamic and probabilistic data, for a classically trained frequentist such as myself will always be grateful.

In addition to these scholarly giants, there are exciting branches of scholarly work currently being flushed out by some of the most dedicated and articulated researchers whom I have ever had the pleasure of coming across. In particular, I am referring to the fMRI studies by Dr. J.R. Anderson, the research on statistics anxiety by Dr. A.J. Onwuegbuzie, or Dr. C. Cavanaugh, Dr. M. Allen, as well as Dr. J.D. Mills for their dedication to understanding online education. Further, none of this work would be possible without the work of scholars who have dedicated much time to identifying, understanding, and assessing statistics cognition; Dr. D. Ben-Zvi, Dr. I. Gal, Dr. C.J. Wild, Dr. M. Pfannkuch, and a special thanks to Dr. R. Beyth-Marom, F. Fidler, and G. Cumming for their call for an evidence-based practice of statistics cognition. While I cannot list every scholarly work that is of importance to the field, I have drafted a list of 10 articles I would recommend to those who are seeking a deeper understanding of statistics education (see Essential Readings for Statistics Education).

As denoted by the quote stated at the beginning of this Chapter, students - especially those in higher education - come to class (for better or worse) with a set of unique cognitive skills, which is largely anchored on how they perceive statistics. By developing a line of research sensitive to how cognition, the environment, affect, and individual differences can play a role in learning, we can enhance the field of statistics education.

Justin Chase
University at Albany (SUNY), USA

REFERENCES

Petocz, P., & Reid, A. (2005). Something strange and useless: Service students conceptions of statistics, learning statistics and using statistics in their future profession. *International Journal of Mathematical Education in Science and Technology*, *36*(7), 789–800. doi:10.1080/00207390500271503

APPENDIX: ESSENTIAL READINGS FOR STATISTICS EDUCATION

Ben-Zvi, D. (2000). Toward understanding the role of technological tools in statistical learning. *Mathematical Thinking and Learning, 2*(1-2), 127–155. doi:10.1207/S15327833MTL0202_6

Ben-Zvi, D., & Garfield, J. (2008). Introducing the emerging discipline of statistics education. *School Science and Mathematics, 108*(8), 355–361. doi:10.1111/j.1949-8594.2008.tb17850.x

Beyth-Marom, R., Fidler, F., & Cumming, G. (2008). Statistical cognition: Towards evidence- based practice in statistics and statistics education. *Statistics Education Research Journal, 7*, 20–39.

Gal, I., & Ograjenšek, I. (2017). Official Statistics and Statistics Education: Bridging the Gap. *Journal of Official Statistics, 33*(1), 79–100. doi:10.1515/jos-2017-0005

Garcia-Santillan, A., Moreno-Garcia, E., Carlos-Castro, J., Zamudio-Abdala, J. H., & Garduno-Trejo, J. (2012). Cognitive, affective and behavioral components that explain attitude toward statistics. *Journal of Mathematics Research, 4*(5), 8–16. doi:10.5539/jmr.v4n5p8

Horton, N. J. (2015). Challenges and opportunities for statistics and statistical education: Looking back, looking forward. *The American Statistician, 69*(2), 138–145. doi:10.1080/00031305.2015.1032435

Lesser, L. M., Wall, A. A., Carver, R. H., Pearl, D. K., Martin, N., Kuiper, S., . . . Weber, J. J. (2013). Using fun in the statistics classroom: An exploratory study of college instructors' hesitations and motivations. *Journal of Statistics Education, 21*, 1–33.

Mills, J. D., & Raju, D. (2011). Teaching statistics online: A decade's review of the literature about what works. *Journal of Statistics Education, 19*, 1–28.

Onwuegbuzie, A. J., & Wilson, V. A. (2003). Statistics Anxiety: Nature, etiology, antecedents, effects, and treatmentsa comprehensive review of the literature. *Teaching in Higher Education, 8*(2), 195–209. doi:10.1080/1356251032000052447

Utts, J. (2015). The many facets of statistics education: 175 years of common themes. *The American Statistician, 69*(2), 100–107. doi:10.1080/00031305.2015.1033981

Acknowledgment

First and foremost, I would like to thank Dr. Zheng Yan for his guidance and encouragement throughout this project.

I would also like to thank all of my mentors: Dr. Jessi L. Smith, Dr. Scott Eidelman, Dr. Diana Newman; and professors: Dr. Joseph Gold, Dr. Shannon McCoy, Dr. Steve Barkan, Dr. Ian Handley, Dr. Heidi Andrade, Dr. Kristi Schilling, Dr. Lisa McAndrew, Dr. Jason Bryer, Dr. Kim Colvin, and Dr. Hung-Bin Sheu. I have learned a unique perspective of statistics and measurement from each of you.

Finally, I would like to thank parents (David and Elizabeth Chase), brother (David Chase II) and to my friends who I thank for support and good conversation; Angela Tampone, Justin LeBreton, Kevin Ruby, Stacey Johnson, Al Avery, Alyssa Pyle, Aaron Paul; and the soon to be doctors Jenny and Jeremiah Martin, Lauren Hawthorne, Abigail Klein-Smith, and Krista Manley.

Justin Chase
University at Albany (SUNY), USA

Acknowledgment

Chapter 1
The Purpose of Statistics Education:
Improving Statistics Cognitions

ABSTRACT

This chapter briefly introduces the field of statistics education, and provides a short synopsis for each of the seven proceeding chapters. In introducing the field, the importance of integrating technology into all aspects of the curriculum, and using a well-designed data driven methodology which takes an interdisciplinary approach, is argued to be central to the success of developing an online statistics course. By integrating technology and using the scientific method to align course materials and adhere to fundamental educational psychological findings, we argue that the primary objective of statistics education—to enhance statistics cognition—can be achieved.

...it is the job—we would argue—of statistical cognition to provide research-based guidance as to how statistical communication can be best accomplished.
- Beyth-Marom, Fidler, & Cuming, 2008, pp. 23-24

1. INTRODUCTION

Much of the research reviewed throughout this book can be traced back to a single question. Namely, what is the nature of schooling? This question,

DOI: 10.4018/978-1-5225-2420-5.ch001

dating as far back to the 1880s (Levin, 1991), would become a central focus of educational discussion as westward expansion continued through the 1930s.

What was at stake was deciding what schools would be designed to do, and there were two wings of education that took opposing stances. The first, led by the grandfather of behaviorism - Edward Thorndike - proposed that the purpose of schooling was to help students obtain a job. That by approaching education with the traditions and values of science and measurement, we could make learning highly effective and efficient (Thorndike, 1918). This mass-education wing was built around the notion that, by using standardized assessments, students learning could be quantified and adjusted to serve the economic needs of the time. Conversely, a progressive reformist named John Dewey advocated for what would become the child-centered wing of education, proposing that the purpose of schooling was to teach children how to become citizens who thought and acted critically for the future we wanted to see.

Ultimately, Thorndike's mass-education wing won this debate in the minds of 20th century education reformists (Lewin, 1991; Tomlinson, 1997). And while the pendulum of educational reform has swung between these two wings, this book proposes a different answer to that solution, one in line with century old discussions framed on the role of statistical scholars in statistics education (see Fox, 1860; Scates & Deming, 1946; Walker, 1945). The nature of education, as far as a scholar in statistics education may be concerned, is how to improve *statistics cognitions* (Beyth-marom, Fidler, & Cumming, 2008).

Statistics cognition can be defined as the process of understanding the mental actions behind remembering, understanding, and applying numerical data. Specifically, scholars have theorized statistics cognition consists of three components (Garfield & Ben-Zvi, 2007); literacy, reasoning, and thinking (Chapter 2).

Having only recently emerged as a unique academic field of its own (Ben-Zvi & Garfield, 2008), integrating statistics cognition into statistics education is no simple task. As will be discussed, and is apparent throughout this book, the literature reviewed is highly interdisciplinary, coming from fields such as psychology, education, economics, engineering, and neurology. Individually, these fields have amassed an impressive literature devoted toward studying how students learn statistics within their respective disciplines.

Thanks in part to the accountability movement (Alexander, 2000; Huisman & Currie, 2004), much research has been centered around educational performance (hereto referred to as performance) as a key educational variable;

with grades being used as the primary means to assess student learning. However, this book focuses on the wide array of educational outcomes, with considerable discussion directed toward cognitive development.

Throughout this book the term *development* is frequently used. This term is used with meaning and distinction, as developmental research has different theoretical approaches (Overton, 2006), that we argue is of greater interest and use to those who care about statistics education. Although there are many nuanced features that set this research apart, critiques have been quick to note that developmental research systematically measures constructs over time (McCall, 1977). More meaningfully, McCall distinguishes between a "can" vs. a "does" question, arguing that developmental scholars need to frame research questions toward the latter.

For instance, while one could ask the question "*can* tests cause statistics anxiety?", the question "*Do* tests cause statistics anxiety?" is more satisfying. Indeed, we already know from an impressive array of literature that statistics anxiety consists of test anxiety (Cruise, Cash, & Bolton, 1985; Onwuegbuzie & Wilson, 2003), with work suggesting that test anxiety is the largest factor explaining statistics anxiety (Onwuegbuzie, 2000). Yet, the fact that tests can cause anxiety is readily apparent to most students. By asking "*do* tests cause statistics anxiety?" we inherently need to appreciate the context through that such anxiety can arise. In order to discuss this context, we are required to study development.

Thus when we talk about *developing* statistics cognitions (Chapter 3), we are primarily talking about improving the ways students understand and use statistical concepts related to their literacy, reasoning, and thinking skills by the end of the term. We argue that statistics educators are primarily interested in improving these skills, and that we must look toward the future of the field to understand how these constructs can be integrated into the statistics courses of tomorrow.

Although the direct path of statistics cognition research is still in its infancy, one element embedded in the future of education is the use of technological advances (Goldin & Katz, 2009). While discussions of the equivalency and rigor of distance education programs has persisted for centuries (Tallent-Runnels et al., 2006; Tawney, 1943), a budding field meshing literature from statistics and online education has started to form (Larwin & Larwin, 2011). Indeed, the role of technology has been cemented into the foundational framework of the field of statistics education (Chance, Ben-Zvi, Garfield, & Medina, 2007).

Although an impressive amount of literature has been generated on teaching statistics online (see Mills & Raju, 2011), covering topics ranging from digital resources (Bialaszewski, 1981; Collis, 1983; Dinov & Christou, 2009; Lesser & Pearl, 2008; Perrett, 2010; Wang, Vaughn, & Liu, 2011), educational assessments (Björnsdóttir, Garfield, & Everson, 2015; Sebastianelli & Tamimi, 2011), course structure (Saadati, Tarmizi, Ayub, & Bakar, 2015; Winquist & Carlson, 2014; Xu, Meyer, & Morgan, 2009), and psychological variables (DeVaney, 2010; Dotterweich & Rochelle, 2012), this book proposes that there is much more to be known that interdisciplinary research has yet to address. It is the hope that with this book, we can provide a more complete understanding of the current nature of online statistics education and viable directions for its future.

2. OVERVIEW

In order to facilitate the use of this book, each Chapter was written with a specific theme in mind. To summarize, we begin by discussing the dimensions and definitions of statistics cognition, highlighting potential areas suited for realignment based in part on observational research that attempts to categorizing cognitions reflected in student outcomes, followed by those that attempted to medically parse neural structures (Chapter 2). Using both forms of research, we propose a neural education model for understanding statistics cognition.

Then, we turn our attention to how this model can be used to inform student learning (Chapter 3) with special emphasis placed on how errors in student learning can arise, and what the sources of these errors mean for the field. Making use of dual process theory (Morewedge & Kahneman, 2009), we outline how student mistakes can be understood, and then address plausible differential student outcomes.

In understanding how student errors can be differentially sourced, we then discuss the role of the digital environment, and what benefits it holds for statistics education (Chapter 4). Centering on a discussion of differentiation, we outline how statistics education can be revamped to suit multiple student needs.

After understanding cognitive structure and online environment, we turn to psychological variables that have been shown to influence cognitive processing, such as student affect (Chapter 5). Throughout our discussion of affect, we highlight the differential role of emotions and attitudes, and discuss what impact they can have for learning statistics online.

Then we discuss the importance of using individual differences in statistics cognition research (Chapter 6), arguing that because an individual difference is nothing more than a means to categorize experience, that we should start looking at cognitively based individual differences. For example, even if men and women were cognitively different, there is no reason to suspect that their statistics cognitions should be inherently different. That is anyone, regardless of gender, should be able to apply the appropriate statistics cognitions as needed. Indeed, we argue for a more dynamic conceptualization of individual differences that even allows one to categorize motivational features as an individual difference.

We conclude this book with a brief synthesis of how each of the Chapters ties together to create a course that is based on statistics cognitions. We then discuss potential future directions of research, as it pertains to the primary topic of each Chapter, to help guide scholars reach a more complete understanding of statistics cognition.

An overarching objective for this book is to connect the distinct areas of statistics education across a variety of fields. In doing so, we hope to have provided a basis for models of statistics cognition to be crafted and assessed. However, the quality of the course constructed is only as good as the evidence that is collected to evaluate student outcomes. As suggested by the quote that started this Chapter, the duty of a statistics educator is to take a data driven approach to enhancing student education by means of developing statistics cognition. As such, it is not only our duty to develop ways to positively influence these cognitive processes, but necessary to record evidence verifying that cognitive growth occurred. As such, the first step in achieving verifiable proof that cognitive development occurred is to inspect the cognitive model of statistics education that is being used.

REFERENCES

Alexander, F. K. (2000). The changing face of accountability: Monitoring and assessing institutional performance in higher education. *The Journal of Higher Education, 71*(4), 411–431. doi:10.2307/2649146

Beyth-Marom, R., Fidler, F., & Cumming, G. (2008). Statistical cognition: Towards evidence- based practice in statistics and statistics education. *Statistics Education Research Journal, 7*, 20–39.

Bialaszewski, D. (1981). Statistics, computer packages, and the real world. *Improving College and University Teaching, 29*(4), 162–165. doi:10.1080/00193089.1981.10533703

Björnsdóttir, A., Garfield, J. B., & Everson, M. (2015). Evaluating two models of collaborative tests in an online introductory statistics course. *Statistics Education Research Journal, 14*, 36–60.

Chance, B., Ben-Zvi, D., Garfield, J. B., & Medina, E. (2007). The Role of Technology in Improving Student Learning of Statistics. *Technology Innovations in Statistics Education Journal, 1*, 1-24. Retrieved from: http://escholarship.org/uc/item/8sd2t4rr

Collis, B. (1983). Teaching descriptive and inferential statistics using a classroom microcomputer. *Mathematics Teacher, 76*, 318–322.

Cruise, R. J., Cash, R. W., & Bolton, D. L. (1985, August). Development and validation of an instrument to measure statistical anxiety. *American Statistical Association Proceedings of the Section on Statistics Education, 4*, 92-97.

DeVaney, T. A. (2010). Anxiety and attitude of graduate students in on-campus vs. online statistics courses. *Journal of Statistics Education, 18*, 1-15. Retrieved from: http://ww2.amstat.org/publications/jse/v18n1/devaney.pdf

Dinov, I. D., & Christou, N. (2009). Statistics online computational resource for education. *Teaching Statistics, 31*(2), 49–51. doi:10.1111/j.1467-9639.2009.00345.x PMID:21297884

Dotterweich, D. P., & Rochelle, C. F. (2012). Online, instructional television and traditional delivery: Student characteristics and success factors in business statistics. *American Journal of Business Education, 5*, 129–137.

Fox, J. J. (1860). On the province of the statistician. *Journal of the Statistical Society of London, 23*(3), 330–336. doi:10.2307/2338500

Garfield, J., & Ben-Zvi, D. (2007). How students learn statistics revisited: A current review of research on teaching and learning statistics. *International Statistical Review*, *75*(3), 372–396. doi:10.1111/j.1751-5823.2007.00029.x

Goldin, C. D., & Katz, L. F. (2009). *The race between education and technology*. Boston, MA: Harvard University Press.

Huisman, J., & Currie, J. (2004). Accountability in higher education: Bridge over troubled water? *Higher Education*, *48*(4), 529–551. doi:10.1023/B:HIGH.0000046725.16936.4c

Larwin, K. H., & Larwin, D. A. (2013). The impact of guided notes on post-secondary student achievement: A meta-analysis. *International Journal of Teaching and Learning in Higher Education*, *25*, 47–58.

Lesser, L. M., & Pearl, D. K. (2008). Functional fun in statistics teaching: Resources, research and recommendations. *Journal of Statistics Education*, *16*, 1–11. Retrieved from https://ww2.amstat.org/publications/jse/v16n3/lesser.pdf PMID:21833159

Levin, R. A. (1991). The debate over schooling: Influences of Dewey and Thorndike. *Childhood Education*, *68*(2), 71–75. doi:10.1080/00094056.1991.10520856

McCall, R. B. (1977). Challenges to a science of developmental psychology. *Child Development*, *48*(2), 333–344. doi:10.2307/1128626

Mills, J. D., & Raju, D. (2011). Teaching statistics online: A decade's review of the literature about what works. *Journal of Statistics Education*, *19*, 1–28. Retrieved from http://ww2.amstat.org/publications/jse/v19n2/mills.pdf

Morewedge, C. K., & Kahneman, D. (2010). Associative processes in intuitive judgment. *Trends in Cognitive Sciences*, *14*(10), 435–440. doi:10.1016/j.tics.2010.07.004 PMID:20696611

Onwuegbuzie, A. J. (2000). Statistics anxiety and the role of self-perceptions. *The Journal of Educational Research*, *93*(5), 323–330. doi:10.1080/00220670009598724

Onwuegbuzie, A. J., & Wilson, V. A. (2003). Statistics Anxiety: Nature, etiology, antecedents, effects, and treatmentsa comprehensive review of the literature. *Teaching in Higher Education*, *8*(2), 195–209. doi:10.1080/1356251032000052447

Overton, W. F. (2006). Developmental psychology: Philosophy, concepts, methodology. In R. M. Lerner (Ed.), *Theoretical models of human development* (Vol. 1, pp. 18–88). New York, NY: Wiley.

Perrett, J. J. (2010). The benefits of using a course disk to aid in the instruction of statistics courses. *Journal of Statistics Education, 18*, 1–21. Retrieved from http://ww2.amstat.org/publications/jse/v18n3/perrett.pdf

Saadati, F., Tarmizi, R. A., Ayub, A. F. M., & Bakar, K. A. (2015). Effect of internet-based cognitive apprenticeship model (i-CAM) on statistics learning among postgraduate students. *PLoS ONE, 10*(7), 1–16. doi:10.1371/journal.pone.0129938 PMID:26132553

Scates, D. E., & Deming, W. E. (1948). Education in Statistics for Participation in Current Affairs. *The School Review, 56*(5), 262–269. doi:10.1086/441508

Sebastianelli, R., & Tamimi, N. (2011). Business statistics and management science online: Teaching strategies and assessment of student learning. *Journal of Education for Business, 86*(6), 317–325. doi:10.1080/08832323.2010.525545

Tallent-Runnels, M. K., Thomas, J. A., Lan, W. Y., Cooper, S., Ahern, T. C., Shaw, S. M., & Liu, X. (2006). Teaching courses online: A review of the research. *Review of Educational Research, 76*(1), 93–135. doi:10.3102/00346543076001093

Tawney, R. H. (1930). The problem of the public schools. *The Political Quarterly, 4*, 117–149.

Tomlinson, S. (1997). Edward Lee Thorndike and John Dewey on the science of education. *Oxford Review of Education, 23*(3), 365–383. doi:10.1080/0305498970230307

Walker, H. (1945). The role of the American Statistical Association. *Journal of the American Statistical Association, 40*(229), 1–10. doi:10.1080/01621459.1945.10500719

Wang, P. Y., Vaughn, B. K., & Liu, M. (2011). The impact of animation interactivity on novices learning of introductory statistics. *Computers & Education, 56*(1), 300–311. doi:10.1016/j.compedu.2010.07.011

Winquist, J. R., & Carlson, K. A. (2014). Flipped statistics class results: Better performance than lecture over one year later. *Journal of Statistics Education*, *22*, 1–10. Retrieved from http://ww2.amstat.org/publications/jse/v22n3/winquist.pdf

Xu, Y. J., Meyer, K. A., & Morgan, D. D. (2009). A mixed-methods assessment of using an online commercial tutoring system to teach introductory statistics. *Journal of Statistics Education*, *17*, 1–17. PMID:21603584

Chapter 2
Understanding Statistics Cognition

ABSTRACT

This chapter focuses on understanding the use of and relationship among the features of statistics cognition: literacy, reasoning, and thinking. We argue that research on statistics cognition is fragmented, which is problematic for understanding how these constructs can be unified to support education. We then review methods of quantifying cognitions, involving studies which have attempted to categorize and parse cognitive processes. This information is then used to synthesize a new approach to understanding statistics cognition, proposing a model which makes specific predictions about the relationship of these features. The model and definitions of cognitions presented in this chapter are used as a basis of discussion cognition throughout the remainder of the book.

Nevertheless, a commitment to testing ourselves—actually seeing how well our predictions work in the real world rather than in the comfort of a statistical model—is probably the best way to accelerate the learning process. - Nate Silver, 2012, p. 292

DOI: 10.4018/978-1-5225-2420-5.ch002

1. INTRODUCTION

In the first Chapter, discussion centered on a simple question, namely, what is the nature of statistics education? Despite its recent emergence as a formalized field (Ben-Zvi & Garfield, 2008), an array of interdisciplinary scholars have produced a vast amount of literature relevant to how students learn statistics (Beyth-marom, Fidler, & Cuming, 2008). However, before we can get into understanding the nature of student learning (Chapter 3), a need exists to understand what is meant by *statistics cognition*. That is, to redefine the components of statistics cognitions (e.g., literacy, reasoning, and thinking), based on an interdisciplinary framework. Such a framework is necessary, as statistical literacy, reasoning, and thinking have distinct scholarly roots, that require an interdisciplinary approach for synthesis of the literature.

Admittedly, what we have outlined is no easy task. Perhaps one of largest obstacles in understanding statistics cognition is to first ask how any type of cognition can be quantified. In fact, for a large portion of history, social scientists demanded directly observable evidence (Benjamin, 2007), often ignoring cognitive processes, or any other construct obscured by the *black box* (Skinner, 1987). Luckily a paradigm shift that emphasizes the importance of cognitive processes has occurred, as even behaviorists are apt to see these schools of thought as more complimentary than adversarial (Beck, 2011). However, to this day, the difficulty of *measuring* and *assessing* cognition remains one of the greatest challenges.

This Chapter serves as a cornerstone of this book, as it provides an in-depth overview of how statistics cognition can be quantified. Specifically, this Chapter begins by defining three central components of statistical cognition: literacy, reasoning and thinking (2.1), and proposes that a need for a new model of statistics cognition exists (2.2). With critiques of statistics cognition in mind, research aimed at categorizing (3.1) and parsing (3.2) cognitions is then reviewed and synthesized in the form of a new model of statistics cognition (3.3). Implications for this model in online environments are discussed (4), followed by a Chapter wide summary (5).

2. PRESENTING STATISTICS COGNITION

In order to elucidate how cognition can be measured, it is first necessary to define our terms. As previously discussed (see section 1), statistics cognition is purported to be comprised of three components: *literacy*, *reasoning*, and *thinking* (Ben-Zvi & Garfield, 2004; delMas, 2002). Traditionally, these concepts have been used in both scientific and colloquial manners for over 100 years. For example, Fox (1860) rigidly used the term "statistical reasoning" to refer to the ability to understand meaning behind statistical information, whereas Dewey (1891) simply used the term to refer to a skill consisting of "right and honest reasoning" (p. 309). While Dewey's interpretation of Kenes argued that statistical reasoning should be no less important than teaching individuals how to "*manage hygiene*", such conceptualizations of key terms varied drastically in their early use.

While historic reviews of statistical literacy (Gal, 2002; 2003; Wallman, 1993; Watson & Callingham, 2002), reasoning (Chan & Ismail, 2014; Garfield, 2003; Garfield & Ben-Zvi, 2009; Ploger & Wilson, 1991), and thinking (Groth, 2006; Hogg, 1991; Pfannkuch & Wild, 2000; Wild & Pfannkuch, 2000) exist elsewhere, the history of these constructs largely denotes that they were developed somewhat independently (see Beyth-Marom et al., 2008). A fascinating occurrence, given that their modern definitions (see Table 1), suggests a high degree of overlap in how these constructs are purported to be related. For example, it would be rather difficult to know the how, when, or why data is analyzed without the ability to explain, question, or compare information (*reasoning*), let alone without an elementary grasp of terms, concepts, and their application (*literacy*).

At face value, it may seem to be little more than a tautological nuance to point out that these constructs developed independently. However, this independence has had important implications for the field of statistics

Table 1. Traditional definition of each dimension of statistical cognition

Dimension	Definition
Literacy (Watson, 1997)	Understanding of terms, language, concepts, and application
Reasoning (Garfield, 2002)	Ability to explain, question, or compare information
Thinking (Wild & Pfannkuch, 1999)	How, when, and why the data is analyzed

education. After all, the diverse roots of these constructs have a significant impact on how we understand the constructs these fields study today. However, in order to understand this fragmentation and its implications for statistics cognition, the core constructs must first be formally defined.

2.1. Statistical Literacy

Originally defined as the ability to evaluate and understand statistical results related to our daily lives (Wallman, 1993), statistical literacy has since evolved to incorporate the ability to discuss opinions regarding statistical information (Gal, 2000) and the ability to interpret key statistical terms and symbols (Garfield & Gal, 1999). Traditionally, statistically literacy can be viewed as having three components; developing understanding of statistical terms; developing understanding of statistical language and concepts; developing the application of statistical techniques to evaluate claims (Watson, 1997).

Although research is converging in a way that suggests statistical literacy is a unique cognitive aspect of statistical education (Rumsey, 2002), comprised of multiple levels (Watson & Callingham, 2003), empirical support for such claims is needed. Regardless as to whether scholars focused on how statistical literacy is needed to understanding graphs (Aoyama & Stephens, 2003), and may relate psychological features such as self-efficacy and interest (Carmichael, Callingham, Hay, & Watson, 2010), the heart of this research has focused on ways in can be improved (Wallman 1993), and how enhancing statistics literacy can positively transform society (Von Roten, 2006).

2.2. Statistical Reasoning

Unlike statistical literacy, which focused on understanding content - such as graphs - and the ability to define terminology and symbols, statistical reasoning is the manner through which individuals make sense of statistical ideas and information (Garfield, 2002). Specifically, statistical reasoning consists of abilities needed to explain, question, or compare statistical information (Rumsey, 2002). Statistical reasoning has distinct roots, embedded in early attempts to examine the underlying heuristics (Nisbett, Krantz, Jepson, & Kunda, 1983) of the decision making process (Kahneman & Tversky, 1982), a topic that has been of great interest to scholars for quite some time (Edwards, 1954). Research has long indicated that everyone, from lay persons (Peterson & Beach, 1967) to statistical experts (Kahneman & Tversky, 1982), seem to be poor *intuitive* statisticians. That is, while we may have a propensity to think that human thought is as logical as the basis for statistical formulas themselves, all too often, people are unable to apply the reasoning of statistical operations to their own thoughts (Kunda & Nisbett, 1986; Lovie & Lovie, 1976).

For example, consider the base rate fallacy (Bar-Hillel, 1980), a cognitive bias of specific information, which generally only pertains to a single case is focused on, while ignoring general information, applies to most cases. For example, consider a scenario where an administrator tells instructors that a new computer program was developed to detect student cheating. This computer program generates a false positive, indicating that a student cheated when they did not, 5% of the time, but the program can never fail to detect work that has truly be plagiarized. Now imagine, that we also know that one in one thousand students will cheat on this assignment based on institutional data. If an educator selects a student at random, and runs their work through this program, and it indicates that the student cheated, how high is the probability that the student really cheated? While many may believe the probability that student cheated is 95%, it is only 2%[1].

Given the failure to commonly apply rationale knowledge to thought (Lambos & Delfabbro, 2007; Oaksford & Charter, 2001), much of the work on statistical reasoning has focused on how it can be improved (Garfield & Ahlgren, 1988; Kahneman, Slovic, & Tversky, 1982). As scholars began to consider how intuitive statistical misnomers could be counteracted (Garfield, 1995; Mevarech, 1983), a new sub-discipline flourished, allowing researchers to determine how statistical reasoning could be assessed (Garfield, 2003) and developed (Garfield & Ben-Zvi, 2009).

Statistical reasoning may often be a central goal of statistical education, because it involves skill processes (i.e., explaining, questioning, and comparing statistical information), and it may be best assessed by using active learning practices (Cobb, 1992). Due to its integration into the judgment and decision making literature (Watkins, Scheaffer, & Cobb, 2010), statistical reasoning may be among the most well researched dimension of statistical cognition (Garfield & Ben-Zvi, 2004). Scholars have generated a variety of models that explicitly look at reasoning (Jones, Langrall, & Mooney, 2004), and have provided a wealth of information on instructional (Chance, Garfield, & delMas, 1999) and assessment (Garfield, 2003; Gal & Garfield, 1997; Watson, 1997) practices. However, the shape of this construct, and its relationship with other the other dimensions of statistics cognition remain somewhat of a mystery (see delMas, 2002).

2.3. Statistical Thinking

Statistical thinking involves understanding how and why data is statistically analyzed (Chance, 2002). Unlike statistical reasoning, that involves combining ideas or connecting concepts to make meaning, statistical thinking generally involves recognizing how, when, and why various analyses should be used. Unfortunately, statistics thinking research is somewhat more sparse relative to the other components; a finding that is not surprising when one considers how the term "*statistical thinking*" has often been used in an interchangeable fashion with "*statistical reasoning*" (see Fong, Krantz, & Nisbett, 1986).

Yet scholars maintain an important distinction among statistical reasoning and thinking (Pfannkuch & Wild, 2000). Statistical thinking is more process than content, suggesting that traditional memory based assessments may not sufficiently measure statistical thinking (see Chapter 3). As a result, a variety of new approaches have been developed by scholars to help educators assess this dimension (Jones et al., 2000; Moore, 2001; Pfannkuch & Wild, 2004; Watson, 1997), with most of these assessments involving active learning techniques (Garfield & Ben-Zvi, 2004). However, as with the previously defined constructs, an independent examination of statistical thinking is informative, but does not afford a complete understanding of statistics cognition.

2.4. Unifying Statistics Cognition

A need for a new cognitive model of statistics education is apparent for at least three reasons. First, a wide degree of fragmentation exists both between and within the field of statistics education, underscoring the need of research synthesis across interdisciplinary academic domains. Second, because models in education are largely assessed by their predictive ability, it is important to consider ways to develop a model that is able to inform how statistics cognitions may be shaped in online settings. Third, a variety of research, which is discussed throughout this Chapter, has approached statistics cognition from both macro and micro-level ends of the empirical spectrum, and so a need exists to connect the cognitive underpinnings that psychologists are interested in, with the educational outcomes that concern educators. To outline why this model is necessary, a brief review of the current state of statistics education is warranted.

2.4.1. The Fragmentation of Statistics Education

Currently, the field of statistics education is said to be fragmented, separating practitioners and researchers (Beyth-Marom, Fidler, & Cuming, 2008). While there have been repeated calls to reform statistics education over the past century (Berkson, 1938; Cummings, 2013; Deming, 1944; Huck, 2007; Garfield, Hogg, Schau, Whittinghill, 2002; Moore, Cobb, Garfield, & Meeker, 1995; Oakes, 1986; Schmidt, 1996; Tishkovskaya & Lancaster, 2012; Tversky & Kahneman, 1982), many have gone unabated. This is not to suggest that researchers are unwilling to practice what their findings suggest, but rather a lack of cohesion exists among findings, making their dissemination and application throughout an interdisciplinary subject, like statistics, exceedingly difficult. For instance, despite the numerous calls to reform assessment in statistics education (see Gal & Garfield, 1997), educational assessment formed as its own field (Bliem & Davinroy, 1997), contributing to a decade long reform movement responsible for reshaping learning objectives and classroom assessment practices as they apply to statistics (Garfield, 1993; Hubbard, 1997).

Similarly, as previously discussed the three theorized features of statistics cognition; literacy (Watson, 1997), reasoning (Garfield, 2002), and thinking (Pfannkuch & Rubick, 2002), have unique etymological roots, and varied scholarly operationalizations (Fong et al., 1986). In fact, our recent literature review conducted, in part for this book, examined 686 peer-reviewed articles relevant to statistics education and found that articles were from highly diverse disciplines (e.g., engineering, psychology, nursing, statistics education), and derived from 501 unique first authors, and published across 110 different scholarly outlets. While such a finding at least reinforce the notion that the field of statistics education is interdisciplinary, the fact that there is little collaboration between scholars of different backgrounds is suggestive of fragmentation.

2.4.2. How Statistics Cognitions Are Used

Generally, statistics education research has noted that developing statistics cognition is a central objective to teaching and learning statistics (Garfield & Ben-Zvi, 2004; 2009). If this is the case, it would suggest that enhancing statistics cognition should result in a higher grade in a statistics course. Namely, that there should be a direct relationship between statistics curriculum and its improvement upon statistics cognition. Even organizations such as the Guidelines for Assessment and Instruction in Statistics Education (GIASE, 2010) seem to dictate that student learning must go beyond memorizing content, and focus on critical thinking and reasoning skills (Hubbard, 1997; Garfield, 1993; Moore, 1997). In fact, such guidelines often suggest an active approach to teaching statistics (Garfield, 1993), as statistics education is primarily concerned with developing a life skill (Ben-Zvi & Garfield, 2004; Gal & Garfield, 1997).

However, as previously noted, often times performance data in terms of course grade is not a highly meaningful variable. For instance, there seems to be little difference in student performance in online and face-to-face classes (see Bernard et al., 2004). Yet, this does not mean that there are no differences worth exploring, as a slew of cognitive and affectual variables still provide important insights in determining how students can succeed based on their educational environments. So, if statistics cognitions may not necessarily predict performance in statistics, how else can we use them?

Because of their vital role to the field of statistics, scholars have primarily studied ways to assess statistics cognitions (Garfield and Ben-Zvi, 2007), as well as educational methods that can be used to enhance statistics literacy (Garfield and Ben-Zvi, 2007), reasoning (Ben-Zvi & Garfield, 2004; Hammerman & Rubin, 2004), or thinking (Makrymichalos, Antony, Antony, & Kumar, 2005; Wild & Pfannkuch, 1999). However, considerably less research has examined what these variables predict (Carmichael et al., 2010), let alone what relationships they may hold with other key variables (Chan, Ismail, & Sumintono, 2013).

Worse yet, what happens when ones statistics curriculum does not result in enhanced statistics cognition? One potential reason why such a discrepancy could arise is because of the gap between educational objectives and the use of content oriented summative assessments (Garfield, 1995; 2002). For example, an educator concerned with measuring whether or not students can define the standard deviation could easily obtain this information using a simple multiple choice or matching exam. However, determining whether or not a student understands why correlation does not imply causation (Aldrich, 1995) is somewhat more difficult. While a question could be crafted to ask a student to select which of the four items best exemplify why correlation does not imply causation, this assessment poses numerous concerns. Ultimately, if educators are concerned with students applying statistical literacy, reasoning, and thinking as a life skill, it is more ecologically valid to prompt students to generate their own rationale, or come to terms with this realization while working first hand with data.

The problem is that objectives geared cognitive development can be difficult to assess, especially in a classroom (Klum, 1990). While education does go beyond assessment, it is imperative to focus on assessment because of its crucial use in evaluating student learning. Without evidence secured by educational assessments (see Chapter 4), it is impossible to determine whether or not targeted cognitions were actually developed. Simply by observing their definitions (see Table 1), one can readily understand why it is so difficult to assess constructs like statistics literacy, reasoning, and thinking. While there has been an urge to shift pedagogical formats from lecturing to active learning (Garfield, 1993), progress may be somewhat marred by the mixed findings regarding the efficacy of whether active learning practices actually improve student outcomes (Carlson & Winquist, 2011; Keeler & Steinhorst, 1995; Weltman & Whiteside, 2010).

Further, fragmentation among scholars is also problematic. As previously discussed, there are a wealth of research pertinent to statistics education, but it is amassed from collecting the works of small pockets of scholars conducting discipline specific research in their own journals. While organizational ties, such as membership in the American Statistical Association may counteract some of this, the problem moves beyond theory and works its way into how distinct practices frame the focus of research. For instance, there may often be different measures and operationalizations for similar constructs depending on the field. Indeed, as previously noted, statistics thinking and reasoning were used in an interchangeable fashion by some scholars.

Yet such fragmentation is not just problematic from a systems perspective. As these different definitions and assessments are used, it makes it exceedingly difficult to collapse across research and make specific predictions about the relationship of statistics cognitions. In essence, this lack of consistency makes it difficult to determine the relationship statistics literacy, reasoning, and thinking have with one another and other variables, such as the environment, motivation and affect.

2.4.3 Theorized Features of Statistics Cognition

Although one may think of the *shape* of a model to be little more than a tautological exercise, the relationship of cognitions is the first step in establishing a realized model of statistics cognition. As depicted in Figure 1,

Figure 1. Theorized models of statistics cognition

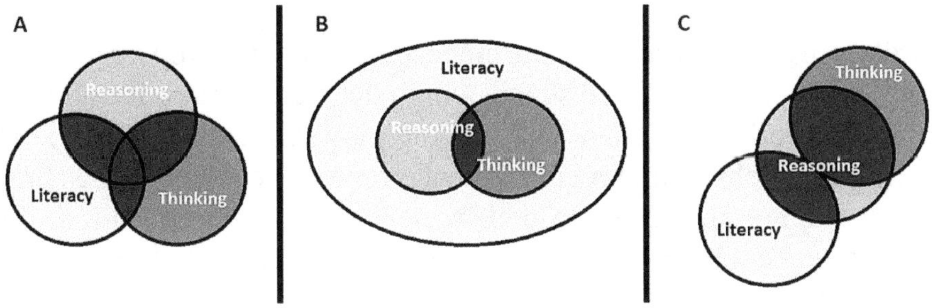

each dimension of statistics cognition (literacy, reasoning, and thinking) could be said to be equally related, overlapping like a three circle Venn diagram (Rumsey, 2002). However, one could also consider statistics cognition to maintain a specific structure, such that one component (statistical literacy) provides a base for the other two dimensions (reasoning and thinking) to operate (delMas, 2002).

As depicted in Figure 1, there are several ways statistics cognitions could be related, with each relationship implying a different assumption in the learning process. Because model A depicts all dimensions of statistical cognition overlapping, it implies that statistical cognition is a live process whereby all dimensions have the ability to be used at once. In terms of learning, this model suggests educators need to create assessments, design courses, and modify instruction so that all three dimensions are scaffolded (Wood, Bruner, & Ross, 1976) conjointly. Conversely, model B suggests literacy provides a base, which is used to enhance the overlapping constructs of thinking and reasoning, suggesting that educators provide a base level of literacy before prompting students to reason and think, so as not to create misconceptions (see Mayer, 2004). Further, one could even imagine a third model, suggesting a hierarchy. As depicted as model C, it may be apparent how a model designed as a linear chain may be more useful for educational prediction, as it suggests a specific progression for cognitive learning. Specifically, this model suggests that educators would provide base level literacy knowledge, and then prompt the student to employ reasoning so that they can constantly re-check their literacy while applying it to their thinking. In essence, there is a need to understand how these cognitions can be studied in an empirical manner.

3. QUANTIFYING COGNITIONS

Regardless of the specific cognitive model employed (Figure 1), a good model should have a clearly defined shape, a high degree of predictive ability, and be useful to both researchers and practitioners (Gigerenzer, 2008). These guidelines will help facilitate a model that is rooted in evidence based practice (Beyth-marom et al., 2008). In their review of statistics education, Beyth-marom and colleagues outline the need to support beliefs, both in the lab and classroom, with scientific findings.

Although calls to treat classrooms as miniature experiments is hardly new (Brown, 1992; Brown, Collins, Duguid, 1989; Cobb, Confrey, diSessa, Lehrer, & Schauble, 2003; Stroud, 1940), Beyth-marom and colleagues present a model of statistics education rooted in cognitive understanding consisting of three elements; normative (what techniques can be used), prescriptive (how to communicate statistical information), and descriptive (how people think about statistics). Because scholars within the field of statistics education may research the same topic, but focus on different sets of theoretical frameworks or empirical questions, it is important to ensure that any model of statistics cognition involves both *qualitative* and *quantitative* findings (Iran-Nejad, McKeachie, & Berliner, 1990), as meaning is lost when psychological constructs are studied in environments with low ecological validity (Bronfenbrenner, 1976; Fang, 1996; Sandoval & Bell, 2004).

To ensure that this model adopts evidence from multiple perspectives, several lines of research will be reviewed. First, systematic qualitative evidence at categorizing cognitions will be reviewed, to provide an understanding of how and why cognitions are assessed in a classroom. Then quantitative work, which focuses on how recent technological advances has helped in scholarly efforts identify cognitive states and begin to parse the cognitions into separate processes rooted in a neurological basis.

3.1. Categorizing Statistics Cognitions

In the age of big data, there is a premium on empirical evidence, that has long ruled scientific findings (see Burman, 1997). However, the need for mix-method research has found a resurgence among a new wave of researchers who are often striving to establish robust results in realistic environments (Creswell, 2002; Cummings, 2013; Johnson, Onwuegbuzie, & Turner, 2007).

One way scholars make use of qualitative data is have experts generate observations based on patterns of data. In the 1950s, prominent educational psychologist Benjamin Bloom did just this. Although primarily interested in assessing curriculum, Bloom and his colleagues would develop one of the first taxonomy of educational objectives that appreciated critical thinking skills, and noted a difference in mental complexity among different types of items (Krathwohl, 2002). For instance, asking a student to define an event is *simpler* than asking them to evaluate it.

Although formed to evaluate curriculum, Bloom's taxonomy provides a framework to understand and connect cognitions to classroom objectives. Indeed, finding results that are replicable in a classroom context are not only important if researchers are to convince educators to adopt findings, but such studies provide a common set of terms for abstract concepts. Even an educational tool as famous as Blooms taxonomy (Bloom, Engelhart, Furst, Hill, & Krathwohl, 1956), was originally ignored because educators where unfamiliar with what a *taxonomy* was (Krathwohl, 2002). By making sure instruments to assess student outcomes are rooted in practice and built from a common tongue, categorizing cognitions can provide valuable educational insight for researchers and educators.

3.1.1. Summary of Bloom's Taxonomy

One way to understand cognition is to consider how cognitions could be framed as learning objectives. Or, in other words, if the goal of the modern statistics class is to develop statistics cognition, then educators must find a way to collect evidence that their methods have done just that. Educators, like scientists, may only make claims that are verified with empirical evidence (Resnick & Resnick, 1992). Thus, to begin to assess cognitive change, educational psychologists developed a framework for categorizing student learning objectives (Bloom et al., 1956).

Initially developed as a means to create a shared test bank among educators of the same discipline (Krathwohl, 2002), Bloom and colleagues sought to find a way that cognitive responses could be assessed (Bloom et al., 1956). However, to Bloom, this taxonomy was far more than a handy assessment device. Bloom wanted to create a means of equating objectives across units, assignments, and curriculums (Krathwohl, Bloom, Masia, 1964).

Bloom's taxonomy is a series of models based on knowledge, emotive, and physiological domains (Bloom et al., 1956), with much of the theoretical framework and empirical work geared toward the knowledge (cognitive) domain (Gronlund, 1998; Seddon, 1978). This bias to measure cognitive domains is likely due to the clear and direct educational application of cognitive dimensions. For example, all educators likely see the need for remembering and understanding information, but it may be more difficult to classify other important behavioral characteristics. For example, Bloom outlines both affective and psychomotor domains, which play a role in measuring and assessing statistics cognition in online environments, but perhaps not as pivotal of a role as cognition. As a result, each of the three domains is briefly described in terms of its function and how it is relevant to this discussion of statistics cognition.

3.1.2. The Cognitive Domain of Bloom's Taxonomy

The cognitive domain of Bloom's Taxonomy is comprised of six distinct categories, each of which containing a series of actions; remember (*recognize, recall*), understand (*interpret, exemplify, classify, summarize, infer, compare, explain*), apply (*execute, implement*), analyze (*differentiate, organize, attribute*), evaluate (*checking, critiquing*), and create (*generating, planning, producing*). Importantly, the actions within each category can be used by educators and researchers to develop a common terminology. For example, according to Bloom's revised taxonomy (Krathwhol, 2002), if a student were asked to "*interpret*" a series of statistical data, they would be answering a question which purportedly assesses their understanding. Conversely, asking a student to recall Bayes that would be a demonstration of a remembered response. Similarly, parallels can be drawn between these sorts of questions, and skills said to fall under the traditional model of statistics cognition (see Table 2).

Already, you may notice that Bloom's taxonomy is at odds end with how statistics cognitions are traditionally defined (see Table 1). For instance, literacy is comprised of understanding statistical language and terms, but this skips the collection and recollection specified in the knowledge level of Bloom's taxonomy. That is, one cannot really understand terms or language until one is able to define, recall, or identify these terms. As such, we propose that these terms be redefined to specify their precise educational objective. That is not to imply that these objectives cannot overlap, but rather, they tend to follow an organized structure based on their mental complexity.

Table 2. Dimensions of statistical cognition arranged by actions

	Statistical Cognition (delMas, 2002)		
	Literacy	**Reasoning**	**Thinking**
Actions	Identify	Why	Apply
	Describe	How	Critique
	Rephrase	Explain	Evaluate
	Translate		Generalize
	Interpret		
	Read		
	Statistical Cognition (Realigned With Bloom's Taxonomy)		
	Literacy (Remember)	**Reasoning (Understand)**	**Thinking (Apply)**
Actions	Recognize	Interpret	Execute
	Recall	Exemplify	Implement
		Classify	
		Summarize	
		Infer	
		Compare	
		Explain	

Specifically, there are two potential benefits of redefining these terms to adhere to Bloom's taxonomy.

The taxonomy forces a clear distinction in educational tasks, therefore providing a clearer means to assess, understand, and ultimately improve the specified dimension of statistical cognition. For instance, if an educator were to administer an exam comprised of 50 questions, each beginning with "*Identify...*", the educator using this overlaid taxonomy would only be providing evidence that student statistical literacy (based in remembering) was assessed. As a result, the educator cannot attest as to whether their instructional methods increased statistics reasoning or thinking. Processes such as reasoning involve carrying out specific tasks to demonstrate one's ability. Indeed, assessing such a cognitively framed goal often requires educators to make use of hands on activities, such as those that involve the inferences about, and execution of, statistical operations (Garfield, 1993; Forbes, Camden, Pihama, Bucknall, Pfannkuch, 2011).

Additionally, Bloom's taxonomy is hierarchical (Bloom et al., 1956; Krathwohl, 2002), suggesting that cognitions must be developed in a specific sequence. For example, in order to interpret, summarize, or infer, one must be able to recognize or recall specific content and procedural knowledge. As such, one must be able to remember information, before they are able to understand it. However, this hierarchical link has been questioned, with some suggesting that while remember, understand, and apply are needed in sequence, analyze, evaluate, and create are less linear and may even be considered cognitive equals (Krathwohl, 2002). While the specific degree of the relationship cannot yet be asserted, there is at least agreement that cognitive processes can be scaffolded to facilitate more complex operations (Belland, 2011; Park & Reuter-Lorenz, 2009; Rogoff, 1998).

Collectively, the cognitive domain of Bloom's taxonomy provides a clear and concise means to label and assess cognition in the classroom. However, scholars must ask whether these the only skills worth assessing from an educational standpoint. Further, while this domain clearly organizes categories of knowledge, a range of motor skills and emotive responses are imperative to developing a quality online statistics courses in higher education.

3.1.3. Affective and Psychomotor Domains of Bloom's Taxonomy

Although the psychomotor and affective domains are considerably less detailed, and have not been the subject of as much scholarly work, they can be thought of as important discussion points for measuring and assessing cognition in online environments. First, affective domains the affective domain of Bloom's taxonomy is designed to gauge attitudes, emotions, and feelings (Krathwohl et al., 1964). Comprised of five hierarchical levels the emotional domain suggests that students may be passively paying attention (receiving), but could also be active participants (responding). Once knowledge is obtained, students often attach meaning to this knowledge (value), which is categorized in schemas (organizing), allowing the student to build abstract knowledge (characterizing). While these levels may lack a high degree of empirical support, the important take away is that affective domains are essential to learning outcomes (see Chapter 5 for a discussion of affect).

Conversely, the psychomotor stage is comprised of no specified levels (Bloom et al., 1956), but theorized to include skills needed to physically utilize or manipulate a tool. Despite a lack of levels proposed by Bloom and his colleagues, other scholars have proposed a variety of psychomotor levels (Dave, 1970; Harrow, 1972). For example, Simpson (1972) proposed six levels of the psychomotor domain (e.g., use of sensory cues to act, readiness to act, ability to carry out guided responses etc). Again, while these levels are more of a point of discussion, one may readily see how mastering fine motor movements is essential in *online* learning environments. Without proficiency in using a mouse (Wood, Willoughby, Rushing, Bechtel, & Gilbert, 2005), or other tasks that could require fine motor movements (Krampe, 2002), students in online classrooms may find themselves at a significant disadvantage in accessing content and learning as effectively (see Schmidt & Lee, 1988).

To some degree, it should not be considered surprising that these variables have received relatively less attention than cognitive and behavioral outcomes. As previously noted, one reason why behavioral outcomes reigned supreme as that they are fairly straight forward and easy to assess. Cognitions are less obvious, with psychomotor and affective variables often existing as an afterthought to the instructor already pressed for time. However, these variables are crucial to at least consider when it comes to online learning.

For example, to teach an online statistics course, a variety of fine motor skills are needed for the student to be successful. As noted, the student must have the coordination to reproduce simple acts, that could be taken for granted. Indeed, professors may often choose to spend time developing learning objectives geared toward statistics literacy and reasoning, neglecting goals that are aimed at ensuring that all students are capable of imitation of content presented in videos (Karppinen, 2005), or manipulation of computerized data in programs such as Excel™ (Warner & Meehan, 2001). By their nature, one could argue that success in online learning is likely related to computer literacy. While some have found that variables such as computer literacy do not influence success (Marcal & Roberts, 2000), others have found such variables to be meaningful indicators (Link & Marz, 2006). At the very least, knowledge of and a plan to account for such student differences will allow an educator to successfully manage their digital course.

While psychomotor control may be best left to instructional designers working to make digital classes more inclusive, the discussion of one's ability to appropriately apply psychomotor control is directly related to affect, which is of central interest in examining differences between online and face-to-face courses. Although there may be few grade based differences in face-to-face and online courses (Bernard et al., 2004), online learning is associated with a stereotype (Benson, 2003), which can vary based on individual differences such as age or experience with technology (Dobbs, Waid, & Carmen, 2009), but can also be influenced by environmental constraints, such as whether the course was required or an elective (Smart & Cappel, 2006). While these differences may not amount to viewable outcomes in terms of a student's grade, they have the potential to shape the atmosphere of the learning environment, and can lead students to different conclusions about a course's usefulness (Song, Singleton, Hill, & Koh, 2004), and have a direct impact on their satisfaction (Richardson & Swan, 2003; Swan, 2001) and motivation (Hoskins & Van Hooff, 2005) for online learning.

While it is beyond the scope of this book to generate specific hypotheses about affect and learning, future research needs to consider these as primary indicators of course quality that can help ensure that online learning is an attraction, as opposed to a repellant. However, in order to achieve this, scholars must do more than attempt to categorize cognitions. There needs to be scientific evidence that demonstrates whether there is a basis for separating these cognitions, or whether categorization is little more than a tautological exercise.

3.2. Parsing Statistics Cognitions

While categorizing research has largely started with considering the skills educators wish to develop, and then retroactively classifying them into groups, experiments from medical (Anderson, 1993; Anderson, Pyke, & Fincham, 2016; Goel & Grafman, 1995) and cognitive (Kotovsky, Hayes, Simon, 1985; Newell & Simon, 1972; Simon, 1978; Sweller, 1988) fields have adopted the microscopic perspective, determine whether cognitions could be parsed. Indeed, studies have employed high (fMRI) - and surprising low (tower of hanoi) - technological paradigms to understand the cognitions involved. By creating clever experimental designs (Simon, 1975) and making use of technological breakthroughs (Anderson, Albert, & Fincham, 2005), empirical evidence has emerged that supports the delineation of distinct cognitions for problem solving.

3.2.1. Problem Solving Paradigms

One way to understand how cognitions could be parsed is to design studies that examine how people can solve problems. Indeed, research has been keen to use a variety of problem solving methods to understand cognitive processes (Anderson, 1993; Bransford, Sherwood, Rieser, 1986; Gick &

Holyoak, 1980; Kotovskyet al., 1985). While examining the merits of these paradigms is beyond the scope of this book, this section provides an overview a classic (Simon, 1975) method that has been adapted to modern paradigms (Anderson, 1990; Robinson & Brewer, 2016) to understand problem solving ability in fields like mathematics (Charles, 1987), and to teach computer science concepts (Marzocchi, 2016). This method, known as the *Tower of Hanoi*, is a classical mathematics puzzle, that could be ideal for understanding problem solving in online statistics environments given that it involves the ability to reason, and consists of an interaction among cognitive processes and motor skills (Anderson et al., 2005).

Originating as a simple mathematical game (Hinz, Klavžar, Milutinović, & Petr, 2013), the Tower of Hanoi presents the puzzle taker the challenge of shifting discs (arranged to one side of the board in ascending order) among three round pegs, following three simple rules: (i) disks can only be moved one at a time; (ii) disks must be moved from the top of the stack; (iii) disks must stay in ascending order. While this paradigm may seem exceedingly simple, it has been a paradigm used to produce robust findings throughout the cognitive literature (Hinz, 1992; Egan & Greeno, 1974; Hayes & Simon, 1974). Relevant to our purposes here, Simon (1975) used the paradigm to delineate among several cognitive strategies participants could use to solve the problem. For example, some participants may adopt a strategy whereby small disks are recursively moved until the task is completed. Alternatively, other participants may use a perceptual strategy, whereby participants focus on the largest disk, and backtrack ways that any smaller disks blocking it could be rearranged. Importantly, Simon suggested that the perceptual strategy is more sophisticated, possibly denoting a more complex set of cognitions used to solve the problem.

The applications of the Tower of Hanoi is as far reaching as the game is simple to play. It has been used to assess frontal lobe (planning) functions by neuropsychologists (Goel & Grafman, 1995; Xu & Corkin, 2001), shown to be related to inhibition and working memory capacity (Welsh, Satterlee-Cartmell, & Stine, 1999), and even be a predictor of fluid intelligence (Zook, Davalos, DeLosh, & Davis, 2004). While the exact nature and rationale behind these connections is widely reviewed elsewhere (see Goel & Grafman, 1995; Humes, Welsh, Retzlaff, & Cookson, 1997; Jurado & Rosselli, 2007; Scholnick, Friedman, & Wallner-Allen, 1997), findings from these studies suggest that there are different methods to solve the Tower of Hanoi, that may help delineate among neurological structures.

3.2.2. fMRI Studies

Because measuring cognitions is difficult, scholars must be exceedingly clever in their experimental methodology. However, cleverness is not enough, and often scholars are limited by the tools at their disposal. Since the 1990s, a brain scanning technique known as *functional magnetic resonance imaging* (fMRI), has been a widely used tool to map the brain (see Amaro & Barker, 2005; Matthews & Jezzard, 2004). By scanning for changes in blood flow, fMRI is thought to measure the energy consumed by brain cells (Poldrack, 2008), thereby allowing inferences about neural activity (Huettel et al., 2009). However, just whether or not - and to what extent - such inferences are warranted, is still under investigation (Jueptner & Weiller, 1995; Heeger & Ress, 2002).

Using the fMRI paradigm, scholars have attempted to identify brain regions in high use during various problem solving activities (Scholnick et al., 1997). For example, Anderson and colleagues trained participants to use a perceptual strategy (Simon, 1975), to solve the Tower of Hanoi. Importantly, this strategy involves a series of rules, such as "...find the largest disk not in its goal position and make the goal to get it in that position" and "[i]f there are any disks blocking the goal move, find the largest blocking disk... and make the new goal to move this blocking disk to the other peg..." (Anderson et al., 2005; p. 9). By teaching participants this strategy, Anderson and colleagues developed an analysis of the cognitive demands for each step, allowing inferences to be made about whether or not the step involves encoding, goal creation, updating information, or retrieval.

Anderson and colleagues interpreted the fMRI results to suggest a negative relationship between the motor and parietal regions during the Tower of Hanoi task (i.e., when intellectual demand was high, motor activity [moving the pegs] was low). Their finding that latency response was a predictor of behavior, suggested to the authors that cognitive planning might be traceable without observation of motor responses.

Putting their theory to the test, Anderson and colleagues (2016) employed the fMRI paradigm to identify and parse the different mental stages needed to solve a mathematics problem. Employing multi-voxel pattern analysis (Mahmoudi, Takerkart, Regragui, Boussaoud, Brovelli, 2012; Norman, Polyn, Detre, & Haxby, 2006; Tong & Pratte, 2012), Anderson and colleagues identified four distinct stages cognitive stages used to solve math problem (i.e., solving, encoding, planning, and responding phases), each corresponding with different neural regions.

Problem solving was found to have the lowest activity, which Anderson and colleagues interpreted to mean that encoding, planning, and responding phases can be viewed "...as adding something to the solving-stage pattern" (p. 7), possibly denoting a problem solving cognitive hierarchy. Further, Anderson and colleagues identified specific neural activity among distinct regions during each of the other three phases. For example, encoding was found to show increased activity in posterior cingulate and polar frontal areas, whereas planning indicated increased activity in ventral visual stream and parietal, prefrontal, and caudate areas. These results, that suggests at least four cognitive processes in solving mathematics problems, mirrored previous findings (Anderson & Fincham, 2014).

Justification for these divisions was, in part, verified by measuring the relationship of neural activity with increased complexity in each of the targeted domains. For example, the neural signature for the planning stage was longer when problems were less obvious to solve. Similarly, the responding stage duration was increased when the complexity of the motor actions required to successfully answer a problem were higher. Additionally, the solving stage duration increased as a function of the number of calculations required to produce an answer. The durations of these increased activity suggests that, to some degree, mathematical problem solving can be compartmentalized into four unique mental stages.

3.3. Synthesizing Approaches: A New Model of Statistics Cognition

As discussed throughout the previous sections, some research attempted to label cognitions based on a post-hoc methodology, framing mental stages in taxonomies that are built upon systematic observations (Krathwohl, 2002). Conversely, others have approached the separation of mental by looking at biological evidence within the individual (Anderson et al., 2016). In one sense, these methods could be considered complementary approaches to studying the same question. The only difference is that with categorization approaches this problem from externally viewable responses in a top-down framework, while parsing cognitions internally employs a bottom-up methodology.

However, these methods are not without their drawbacks. As discussed at the beginning of this Chapter, quantifying cognitions is an incredibly difficult task, and imposing a logical rationale onto responses after they have been generated, could lead scholars astray. It is tempting to impose logical structures and taxonomies on observations because it allows educators to act and understand how cognitions can be assessed in a classroom context (Bloom et al., 1956). However, we must remain vigilant, for our cognitions - much like anything in our universe - are under no obligation to make sense to us. That is, while we may impose frameworks onto our cognitions, these frameworks may not be reflective of reality.

Similarly, we must be careful to avoid using subtractive reasoning (Knauff, Fangmeier, Ruff, & Johnson-Laird, 2003; Newman, Twieg, & Carpenter, 2001) and remain critical of the data based inferences we make (Brett, Johnsrude, & Owen, 2002; Stark & Squire, 2001; Vul, Harris, Winkielman, & Pashler, 2009). To illustrate the ease of erroneous findings in fMRI studies, researchers loaded a dead Atlantic Salmon into an fMRI, and administered an open-ended *mentalizing* task, whereby the dead fish was exposed to photographs depicting social interactions of different emotional valence (Bennett, Baird, Miller, & Wolford, 2009). When not controlling for multiple comparisons, a variety of significant differences emerge, that could otherwise suggest the fish was engaged in perspective-taking. While humorous, this study illustrates the danger of relying entirely on statistical analysis to drive theory, and reaffirms the need for context in analyzing data.

3.3.1. Neural Education Model of Statistics Cognition

Statistics education is a widely diverse field, that has recently emerged as a unique discipline (Ben-Zvi & Garfield, 2008). However, the interdisciplinary nature of statistics education is both a blessing and curse. A blessing in that a wide array of relevant literature using both quantitative and qualitative methodologies exist, that will be imperative for understanding the context for relevant empirical findings (Creswell & Clark, 2007). A curse in that the sheer number of different scholars and publishing highly meaningful research is vast, and difficult to synthesize. The previously outlined branches of research (categorizing and parsing) represent two targeted methods to understanding the same construct: statistical cognition. This method was employed because, as previously discussed, the findings throughout the interdisciplinary field of statistics education are fragmented (Beyth-Marom et al., 2008). By employing both top-down (categorizing) and bottom-up (parsing) frameworks, a more complete understanding of statistical cognition can be realized.

Notes: General cognitions are listed as observable. Given the framework discussed throughout this Chapter (see Anderson et al., 2016), an emerging body of evidence is building, supporting the validity of fMRI cognitive parsing. Further, the keen observe may note that motor responses are missing from this model. While understanding the cognitions behind motor responses is crucial to assessing and measuring statistics cognition, it remains unclear how this construct is related, given that considerably less research has focused on understanding the nature of the relationship among these dimension. In synthesizing the literature reviewed throughout this Chapter, a three-tiered model of statistics cognition, labeled the *neural educational model* of statistics education, is proposed (Table 3). This model outlines a framework that educators and scholars can use to communicate and examine similar constructs across their respective domains. As discussed throughout this book, this model is intended to provide a basis to test specific empirical relationships among statistics cognition, and a variety of educational psychological constructs.

Specifically, this model allows researchers and educators to share common ground by equating general cognitions with educational objectives via statistics cognitions. For example, an educator who notices students who have had difficulty in tasks requiring them to remember information, may note a weakness of statistical literacy, that could be due to an inability to encode information. Importantly, the model presented within this book is not meant to be exhaustive, as additional actions for educational objectives can be observed (Table 2). Because this model is not exhaustive, there could be additional cognitions worth considering. For instance, while the study carried out by Anderson and colleagues (2016) indicates a specific four stage pattern of solving mathematical problems, statistical problems may be inherently different, requiring similar cognitions at different frequencies, or perhaps different cognitions altogether.

Table 3. Educator key words for the neural education model of statistics cognition

	Domain	Dimension	Action(s)	Cognition(s)
Statistics Cognitions	Statistical Literacy	Remember	Recognize, Recall	Encoding
	Statistical Reasoning	Understand	Interpret, Exemplify, Classify, Summarize, Infer, Compare, Explain	Planning
	Statistical Thinking	Apply	Execute, Implement	Solving
Statistics Metacognitions	Generalized	Analyze	Differentiate, Organize, Attribute	Multiple
	Generalized	Evaluate	Examining, Critiquing	Multiple
	Generalized	Create	Generating, Producing	Multiple

Further, the specific problem solving situation that participants are exposed to in much of laboratory research may pose other constraints that reduce ecological validity of the effects in online statistical environments. Indeed, the reviewed findings do not allow inferences to be made about how cognitions may differ if a student were working on a take home statistics project, as opposed to a test involving arbitrary calculations. In such a project the role of retrieval is of critical importance relative to the role of encoding. However, it would be a difficult claim to state that mental retrieval of specific information could occur without encoding of said information.

This model is not so much about considering what other cognitions and situations are of psychological importance. In fact, the evidence that does exist has been incorporated into this model. For instance, Anderson and colleagues (2016) found that encoding and planning adding to brain activity related to problem solving. As a result, planning and encoding can impact solving, but solving does not appear to share a bidirectional relationship with the other two constructs. Similarly, this may provide insight into the shape of statistics cognition (Rumsey, 2002; delMas, 2002; section 3.3). Specifically, this model suggests that when working on a statistics problem, students may have a set of general problem solving cognitions active. As they work on the problem, they consider the content they have learned (encoding) and plan their approach (reasoning).

Instead, this model is concerned with linking cognitive stages to educational outcomes, in online environments. In the following sections, the NE model will be discussed in terms of its assumptions, and intended uses in online statistical environments.

3.3.2. Assumptions Within the Neural Education Model

The NE model makes four primary assumptions. (1) This model assumes that there are general cognitions related to one's ability to solve problems, and that, in order to solve problems, individuals rely on the information they have encoded, as well as their ability to reason the appropriate statistical approach. (2) General cognitions influence what we label domain specific, or statistical cognitions, that impact educational outcomes. (3) Cognitions can be used to a specific end, that we have categorized as *meta-cognitions*; analyze, evaluate, create, all involve using the base levels of statistics cognitions (literacy, reasoning, and thinking) to successfully carry out a statistical task (Table 3). (4) An invariant hierarchy exists among cognitions, but is altered in terms of education objectives. While encoding and planning are said to impact solving, remembering and understanding may not necessarily influence applying these skills. That is, application of skills is built upon remembering that is built upon understanding.

Additionally, a series of implicit assumptions could also be considered regarding this model. For instance, this model implies that statistics literacy and reasoning are important parts of statistics thinking. However, this model also makes a clear distinction between general and domain specific cognitions. While reasoning may play a general role in problem solving and academic performance, this model would consider "general reasoning" to be a parent component of statistics reasoning. Indeed, some informal evidence may exist for the proposed structure of statistics cognition. For instance, research on transfer of learning suggests that students have a difficult time taking a skill learned in one domain, and applying it to another (Thorndike & Woodsworth, 1901; Haskell, 2000). Thus, it stands to reason that students have developed an array of problem solving skills by the time they reach higher education, but that they are honing those general skills for domain specific tasks. Because this information is stored in terms of its domain, as opposed to a general skill, students could experience difficulty learning statistics in courses that treat literacy, reasoning, and thinking as constructs to be individually introduced and assessed.

3.3.3. Re-Presenting Statistics Cognition

As depicted in Table 3 and Figure 2, the NE model of statistics cognition may seem to redefine the core constructs; literacy, reasoning, and thinking so that they are retrofitted into Bloom's revised taxonomy (Krathwohl, 2002). However, this model is based less on a new definition, and more on re-categorizing the educational components theorized by previous definitions. For example, delMas (2002) argues that statistical literacy involves identifying information, which as per the neural educational (NE) model, is also part of the concept of literacy. However, unlike the model proposed by delMas, which suggests that describing involves and interpreting are also part of literacy, the NE model proposes that proposes a shift in these terms based on research that has categorized and parsed cognitions.

For instance, as presented in Table 1, literacy refers to understanding terms, language, concepts, and their application. In one sense, literacy can be viewed as the ability to encode and retrieve information relating to statistical

Figure 2. Theorized cognitive path of the neural education model of statistics cognition

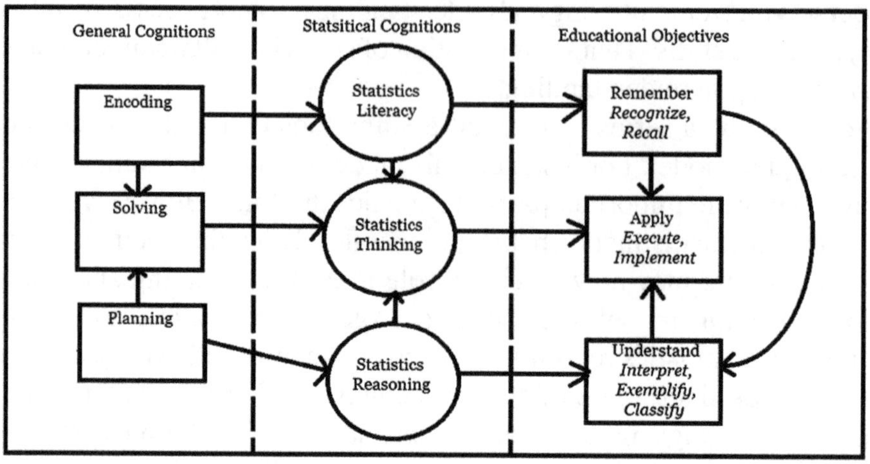

language. While literacy can involve understanding what terms mean, and how they are applied (Watson, 1997), we propose a narrow interpretation of this statement, such that we encode and retrieve the processes used to apply information, and that the conceptual knowledge portion of understanding is literacy.

While literacy may involve encoding and decoding, planning involves executive resources (Owen, 1997), and the ability to think ahead. Planning is primarily concerned with how to achieve one's goal (see Hoc, 1988; Das, Kar, Parrila, 1996). As a result, parallels between reasoning and planning are suggested. While other models, such as the one proposed by delMas (2002) suggest that statistics reasoning involves explaining how and why something occurred, the NE model suggests that planning involves the actual thinking behind how and why something was done. In an educational context, we often may assess this dimension by asking students to showing their work, and explaining their reasoning, but the behavioral component of explaining actually involves statistical thinking, according to the NE model.

Statistics thinking involves the application side of statistics, namely how students go about carrying out actions to achieve their goals. It involves taking what they know from the content they have learned (literacy) as well as their ability to think about how and why they should carry out a certain procedure (reasoning), and applying it to statistical problems (thinking). In line with evidence presented in Anderson (2016), statistics thinking may be an online system activated for the duration of the problem solving attempt.

As students work with their statistical problems, they then direct efforts to make use of specific content guided by their literacy and reasoning.

3.3.4. Domain-Specific and Generalized Cognitions

One of the more obvious implications of the NE model of statistics education (Figure 2) is that it delineates among domain specific (statistics) and general cognitions. That is, it is important to acknowledge that an array of general cognitions may exist within an individual (Salomon, 1997; Steinberg, 1989), but with statistics education, these general cognitions become pruned, and are directly applied to the course subject-matter. Such a claim is parallel with long standing cognitive research that theorized that reasoning (Gigerenzer & Hug, 1992), conceptual change (Carey & Spelke, 1994), and even strategic knowledge (Alexander & Judy, 1988) may all be partially domain specific. Although it is currently beyond the scope of the NE model to specify the mechanism for this cognitive refinement, it is important to acknowledge the difference between general and domain specific cognitions from both an educational and scholarly perspective.

One primary reason as to why such a distinction is meaningful is that researchers have suggested people have difficulty applying knowledge and skills learned in one subject to another (Perkins & Salomon, 1992). This body of research, known as *transfer of learning*, is a psychological line of study examining how experiences influence the ability to learn information. Although a wide variety of ways to classify various types of transfer exist (Haskell, 2000), for the purpose of statistics cognition, transfer has to do with whether or not a general cognition can be transferred to statistics cognitions. As outlined in the NE model (Figure 2), a set of general cognitions, having to do with one's ability to encode, retrieve, plan, and problem solve exist. However, these skills are then honed in terms of domain specificity. That is, the learner uses their general cognitive processes and applies it to their statistical information.

This is, in effect, the primary purpose of statistics education. To help students take their cognitive abilities and enhance them for domain specific purposes. Such a distinction is important, as it implies that statistics cognition, while linked to base level cognitive facilities at a neurological level, is more of a byproduct of effortful focus within an appropriate statistical curriculum (see Chapter 4).

Further, this delineation among statistics and general cognitions is important because it accounts for domain level perspectives. For instance, it suggests that you could have a student who is extremely gifted, and has a high capacity for problem solving by means of planning, encoding, retrieving, and applying the appropriate skills. However, external features, such as the environment (Chapter 4), affect (Chapter 5) and perhaps even individual differences (Chapter 6) can influence the development of a cognitive structure. This could be taken as evidence to suggest what would be labeled *transfer of cognitions*. That is, the way a cognition is refined in one specific domain may not necessarily transfer to another domain.

4. ONLINE LEARNING IMPLICATIONS: AN APPLICATION OF THE NE MODEL IN ONLINE ENVIRONMENTS

In order to have a meaningful discussion of online statistics learning, we must consider an online course that is framed on developing statistics cognition. In essence, we are striving to develop digital courses that enhance student statistics literacy, reasoning, and thinking by the end of the term. However, as outlined throughout this Chapter, there is a relative lack of cohesion among the study of cognitions as they apply to statistics education, and so a framework making use of interdisciplinary work was offered as a point of reference for the work outlined in this book. Although the focus has been on a model of statistics education, we acknowledge that online learning is the future of this emerging field. Indeed, as online and distance education deliver viable solutions to the serious geographical and monetary barriers many students face (see Harasim, 1996), statistics education will continue to integrate technology (see Chance, Ben-Zvi, Garfield, & Medina, 2007; Larwin & Larwin, 2011; Mills & Raju, 2011; Mills & Xu, 2006).

In one sense, a quick examination of the literature reveals that there are relatively few differences among online and face-to-face courses. For instance, in a meta-analysis of 242 studies, there were no meaningful differences in achievement, attitude, or retention (Bernard et al., 2004). In a sense, this is to be expected, as it suggests that online courses can be just as good as face-to-face learning environments.

However, there may be a bias in terms of how these concepts are studied throughout the experiments included in these meta-analyses. For example, most research likely focuses on student grades as a means of course level achievement. However, constructs such as GPA alone are not enough to build highly predictive models, as psychological variables provide considerable context (Conrad, 2006). This also should not be too surprising given that a student's GPA is computed from each individual course, which may have different forms of assessments, or even different types of learning objectives (Chapter 4). If statistics courses are supposed to develop cognition, this itself suggests that a grade in a statistics course should represent their cognitive - as opposed to behavioral - abilities. Again, this is not to suggest that courses must adopt either cognitive or behavioral learning objectives, but rather suggesting that differences in the frequency of objectives that a course adopts could influence exactly what that grade is predictive of.

Predictive or not, course grades provide a valuable means to describe a student's performance in a quick and easy manner. While there may be no significant differences comparing student outcomes between learning modalities, the scope of studies included in such analyses are often diverse, which some have suggested may wash out any significant effects (Zhao, Lei, Yan, Lai & Tan, 2005). Although there may be temptation to treat a digital environment as inherently different from a face-to-face learning, there are a wide range of curricular factors that can influence the quality of a program, regardless of its modality.

For instance, in their meta-analysis of 51 studies, Zhao and colleagues (2005) found the factors that improve the quality of distance education, such as the types of interactions presented to students, also improve the quality of face-to-face learning. However, there seemed to be an additive effect among students exposed to both face-to-face and online learning, leading Zhao and colleagues to conclude that there needs to be an appropriate mixture of human contact with technology.

Although differences among things such as prior experience with technology and the discipline of the content were observed, such differences could be easily explained. For example, Zhao and colleagues found that computer science seemed to be more effective when taught online, but there is no way to disentangle whether self-selection bias of online students influenced the findings. Similarly, while undergraduate students seemed to benefit more from online learning than graduate students, Zhao and colleagues readily acknowledge that undergraduate courses may focus more on knowledge and understanding domains of learning, as opposed to graduate courses that may push students to use critical thinking more frequently. This means that it should be just as plausible to design a course that enhances statistics cognition online, as it is to build one in a face-to-face learning environment.

Fundamentally, this changes the question addressed by this book from how to develop an online statistics course, to how to develop a quality course in a way that can be effectively and efficiently administered in a digital environment (see Chapter 4). Luckily for educational scholars, it seems apparent that good education is simply good education, regardless of its modality (Shale, 1990). While the quality of digital programs can vary widely, so can programs that primarily offer face-to-face learning. Given that modality is not a direct indicator of difference in student achievement or affect, it makes little sense to compare online with face-to-face learning environments. Instead, this book will focus on the guidelines offered by educational research that can be applied to online learning environments. Although their application may seem more difficult in a digital environment, scholars have amassed excellent work outlining specific strategies for developing quality online courses (see Anderson, 2008; Dabbagh & Bannan-Ritland, 2005; Swan, 2001).

5. SUMMARY: APPLYING A SYNTHESIZED MODEL OF STATISTICS COGNITION

In order to understand how we can teach statistics online, we must first understand what our courses are supposed to do. Because scholarly work largely advocates for statistics educators to develop statistics cognitions (see Ben-Zvi, 2000; Garfield & Ben-Zvi, 2007; Garfield, 2002), it means that statistics outcomes must be linked to cognitive processes. A task that becomes even more complicated when considering the unique historical roots each statistics cognition is derived from. Thus, we proposed a new model to understand statistics from cognitions to educational outcomes so that we could consider ways to design an online course around these objectives. Specifically, we present our model to address the fragmentation of the field, while describing its predictive and explanatory relationships within itself.

To be of use, such a model must encompass evidence based practices aimed at synthesizing qualitative and quantitative information. By considering how statistics cognition can be categorized as educational objectives, and by identifying distinct cognitive patterns, a variety of methods can be used to approach measuring statistics cognition. After all, if the field of statistics cognition is, indeed, fragmented, the only hope of a true discovery is to triangulate systematic approaches, allowing educators and scholars to measure and assess statistics cognition uniformly.

In proposing the neural education (NE) model, it is hoped that the calls for a comprehensive reformation of statistics cognition based on multiple theoretical perspectives is taken seriously. Indeed, it is our hope that the NE provides a basis to make specific predictions about important educational psychological constructs in higher education online environments. It is not meant to be exhaustive, nor it is meant to suggest that this is the "only" model that could be used to explain statistics cognition. In fact, this model draws from specific branches of research, that in of themselves have unique strengths and weaknesses based on their own interdisciplinary roots. However, it is established as a spring board to understand psychological variables in online statistics environments.

To illustrate the value of this model, it will be briefly discussed in the context of student learning, environmental and curriculum, affective processes, and individual differences. Specifically, while this Chapter provided the framework, the rest of the book aims to illustrate the utility of the model to both researchers and educators. Specifically, this model is used to assess how cognitions develop as a function of student learning (Chapter 3), what environmental features influence statistics cognition (Chapter 4), and an in-depth discussion of how characteristics such as affect (Chapter 5) and individual differences (Chapter 6) can impact educational outcomes that are framed upon student cognition. In the traditional positivist fashion, this model represents an attempt to stand on the shoulders of giants and provide a forum of discussion for the future of statistics education.

REFERENCES

Aldrich, J. (1995). Correlations genuine and spurious in Pearson and Yule. *Statistical Science*, *10*, 364–376.

Alexander, P. A., & Judy, J. E. (1988). The interaction of domain-specific and strategic knowledge in academic performance. *Review of Educational Research*, *58*(4), 375–404. doi:10.3102/00346543058004375

Amaro, E. Jr, & Barker, G. J. (2006). Study design in fMRI: Basic principles. *Brain and Cognition*, *60*(3), 220–232. doi:10.1016/j.bandc.2005.11.009 PMID:16427175

Anderson, J. R. (1990). *Cognitive psychology and its implications* (3rd ed.). New York, NY: W.H. Freeman/Times Books/ Henry Holt & Co.

Anderson, J. R. (1993). Problem solving and learning. *The American Psychologist, 48*(1), 35–44. doi:10.1037/0003-066X.48.1.35

Anderson, J. R., Albert, M. V., & Fincham, J. M. (2005). Tracing problem solving in real time: fMRI analysis of the subject-paced Tower of Hanoi. *Journal of Cognitive Neuroscience, 17*(8), 1261–1274. doi:10.1162/0898929055002427 PMID:16197682

Anderson, J. R., & Fincham, J. M. (2014). Discovering the sequential structure of thought. *Cognitive Science, 38*(2), 322–352. doi:10.1111/cogs.12068 PMID:23941168

Anderson, J. R., Pyke, A. A., & Fincham, J. M. (2016). Hidden stages of cognition revealed in patterns of brain activation. *Psychological Science, 27*(9), 1215–1226. doi:10.1177/0956797616654912 PMID:27440808

Anderson, T. (2008). *The theory and practice of online learning*. Alberta, Canada: Athabasca University Press.

Aoyama, K., & Stephens, M. (2003). Graph interpretation aspects of statistical literacy: A Japanese perspective. *Mathematics Education Research Journal, 15*(3), 207–225. doi:10.1007/BF03217380

Bar-Hillel, M. (1980). The base-rate fallacy in probability judgments. *Acta Psychologica, 44*(3), 211–233. doi:10.1016/0001-6918(80)90046-3

Beck, J. S. (2011). *Cognitive behavior therapy: Basics and beyond.* New York, NY: Guilford Press.

Belland, B. R. (2011). Distributed cognition as a lens to understand the effects of scaffolds: The role of transfer of responsibility. *Educational Psychology Review, 23*(4), 577–600. doi:10.1007/s10648-011-9176-5

Ben-Zvi, D. (2000). Toward understanding the role of technological tools in statistical learning. *Mathematical Thinking and Learning, 2*(1-2), 127–155. doi:10.1207/S15327833MTL0202_6

Ben-Zvi, D., & Garfield, J. (2008). Introducing the emerging discipline of statistics education. *School Science and Mathematics, 108*(8), 355–361. doi:10.1111/j.1949-8594.2008.tb17850.x

Ben-Zvi, D., & Garfield, J. B. (2004). Introduction to statistical literacy, reasoning, and thinking. In D. Ben-Zvi & J. B. Garfield (Eds.), *The challenge of developing statistical literacy, reasoning and thinking* (pp. 1–17). Dordrecht, Netherlands: Kluwer Academic Publishers. doi:10.1007/1-4020-2278-6

Benjamin, L. T. (2007). *A brief history of modern psychology*. Oxford, UK: Blackwell Publishing.

Bennett, C. M., Baird, A. A., Miller, M. B., & Wolford, G. L. (2009). Neural correlates of interspecies perspective taking in the post-mortem Atlantic salmon: an argument for proper multiple comparisons correction. *15th Annual Meeting of the Organization for Human Brain Mapping*, San Francisco, CA. doi:10.1016/S1053-8119(09)71202-9

Benson, A. D. (2003). Assessing participant learning in online environments. *New Directions for Adult and Continuing Education, 100*(100), 69–78. doi:10.1002/ace.120

Berkson, J. (1938). Some difficulties of interpretation encountered in the application of the chi- square test. *Journal of the American Statistical Association, 33*(203), 526–536. doi:10.1080/01621459.1938.10502329

Bernard, R. M., Abrami, P. C., Lou, A. Y., Borokhovski, A. W., Wade, L., Wozney, L., & Huang, B. et al. (2004). How does distance education compare with classroom instruction? A meta-analysis of the empirical literature. *Review of Educational Research, 74*(3), 379–439. doi:10.3102/00346543074003379

Beyth-Marom, R., Fidler, F., & Cumming, G. (2008). Statistical cognition: Towards evidence-based practice in statistics and statistics education. *Statistics Education Research Journal, 7*, 20–39.

Bliem, C. L., & Davinroy, K. (1997). *Teachers' beliefs about assessment and instruction in literacy. National Center for Research on Evaluation, Standards, and Student Testing (CRESST)*. Los Angeles, CA: Graduate School of Education & Information Studies, University of California. Retrieved from https://pdfs.semanticscholar.org/21c0/2b477b53a7ce3409dee7f63a44613b46bdd5.pdf

Bloom, B. S., Engelhart, M. D., Furst, E. J., Hill, W. H., & Krathwohl, D. R. (Eds.). (1956). *Taxonomy of Educational Objectives, Handbook I: The Cognitive Domain*. New York, NY: David McKay Company Inc.

Bransford, J., Sherwood, R., Vye, N., & Rieser, J. (1986). Teaching thinking and problem solving: Research foundations. *The American Psychologist*, *41*, 1078–1089. doi:10.1037/0003-066X.41.10.1078

Brett, M., Johnsrude, I. S., & Owen, A. M. (2002). The problem of functional localization in the human brain. *Nature Reviews. Neuroscience*, *3*(3), 243–249. doi:10.1038/nrn756 PMID:11994756

Bronfenbrenner, U. (1976). The experimental ecology of education. *Educational Researcher*, *5*, 5–15.

Brown, A. L. (1992). Design experiments: Theoretical and methodological challenges in creating complex interventions in classroom settings. *Journal of the Learning Sciences*, *2*(2), 141–178. doi:10.1207/s15327809jls0202_2

Brown, J. S., Collins, A., & Duguid, P. (1989). Situated cognition and the culture of learning. *Educational Researcher*, *18*(1), 32–42. doi:10.3102/0013189X018001032

Burman, E. (1997). Minding the gap: Positivism, psychology, and the politics of qualitative methods. *The Journal of Social Issues*, *53*(4), 785–801. doi:10.1111/j.1540-4560.1997.tb02461.x

Carey, S., & Spelke, E. (1994). Domain-specific knowledge and conceptual change. In L. A. Hirschfeld & S. A. Gelman (Eds.), *Mapping the mind: Domain specificity in cognition and culture* (pp. 169–200). Cambridge, UK: Cambridge University Press. doi:10.1017/CBO9780511752902.008

Carlson, K. A., & Winquist, J. R. (2011). Evaluating an active learning approach to teaching introductory statistics: A classroom workbook approach. *Journal of Statistics Education*, *19*, 1–23. Retrieved from https://ww2.amstat.org/publications/jse/v19n1/carlson.pdf

Carmichael, C., Callingham, R., Hay, I., & Watson, J. (2010). Statistical Literacy in the Middle School: The Relationship between Interest, Self-Efficacy and Prior Mathematics Achievement. *Australian Journal of Educational & Developmental Psychology*, *10*, 83–93.

Chan, S. W., & Ismail, Z. (2014). Developing statistical reasoning assessment instrument for high school students in descriptive statistics. *Procedia: Social and Behavioral Sciences*, *116*, 4338–4343. doi:10.1016/j.sbspro.2014.01.943

Chan, S. W., Ismail, Z., & Sumintono, B. (2014). A Rasch Model Analysis on Secondary Students Statistical Reasoning Ability in Descriptive Statistics. *Procedia: Social and Behavioral Sciences*, *129*, 133–139. doi:10.1016/j.sbspro.2014.03.658

Chance, B., Ben-Zvi, D., Garfield, J. B., & Medina, E. (2007). The role of technology in improving student learning of statistics. *Technology Innovations in Statistics Education*, *1*, 1–23.

Chance, B., Garfield, J., & delMas, B. (1999). *A model of classroom research in action: Developing simulation activities to improve students' statistical reasoning*. 52nd Session of the International Statistical Institute, Helsinki, Finland.

Charles, R. (1987). *How to evaluate progress in problem solving*. Reston, VA: National Council of Teachers of Mathematics.

Chater, N., & Oaksford, M. (2001). Human rationality and the psychology of reasoning: Where do we go from here?. *British Journal of Psychology*, *92*(1), 193–216. doi:10.1348/000712601162031

Cobb, G.W. (1992). Teaching statistics. *Heeding the call for change: Suggestions for curricular action, 22*, 3-43.

Cobb, P., Confrey, J., DiSessa, A., Lehrer, R., & Schauble, L. (2003). Design experiments in educational research. *Educational Researcher*, *32*(1), 9–13. doi:10.3102/0013189X032001009

Conrad, C. F., & Serlin, R. C. (Eds.). (2006). *The SAGE handbook for research in education: Engaging ideas and enriching inquiry*. Thousand Oaks, CA: SAGE Publications. doi:10.4135/9781412976039

Creswell, J. W. (2002). *Educational research: Planning, conducting, and evaluating quantitative*. Upper Saddle River, NJ: Prentice Hall.

Creswell, J. W., & Clark, V. L. P. (2007). Designing and conducting mixed methods research. *Australian and New Zealand Journal of Public Health*, *31*(4), 388–389. doi:10.1111/j.1753-6405.2007.00096.x

Cumming, G. (2013). *Understanding the new statistics: Effect sizes, confidence intervals, and meta-analysis. Abingdon-on-Thames*. Routledge.

Cumming, G. (2014). The new statistics: Why and how. *Psychological Science*, *25*(1), 7–29. doi:10.1177/0956797613504966 PMID:24220629

Dabbagh, N., & Bannan-Ritland, B. (2005). *Online learning: Concepts, strategies, and application*. Upper Saddle River, NJ: Prentice Hall.

Das, J. P., Kar, B. C., & Parrila, R. K. (1996). *Cognitive planning: The psychological basis of intelligent behavior*. Thousand Oaks, CA: Sage Publications, Inc.

Dave, R. H. (1970). Psychomotor levels. In R. J. Armstrong (Ed.), *Developing and writing educational objectives* (pp. 33–34). Tucson, AZ: Educational Innovators Press.

delMas, R. C. (2002). Statistical literacy, reasoning, and learning: A commentary. *Journal of Statistics Education*, *10*, 1–18.

Deming, W. E. (1944). On errors in surveys. *American Sociological Review*, *9*(4), 359–369. doi:10.2307/2085979

Dewey, D. R., & Keynes, J. N. (1891). Review: Keynes on statistics: The scope and method of political economy by John Neville Keynes. *Publications of the American Statistical Association*, *2*(14), 308–310. doi:10.2307/2276579

Dobbs, R. R., Waid, C. A., & del Carmen, A. (2009). Students' perceptions of online courses: The effect of online course experience. *Quarterly Review of Distance Education*, *10*, 9–26.

Edwards, W. (1954). The theory of decision making. *Psychological Bulletin*, *51*(4), 380–417. doi:10.1037/h0053870 PMID:13177802

Egan, D. E., & Greeno, J. G. (1974). Theory of rule induction: Knowledge acquired in concept learning, serial pattern learning, and problem solving. In L. W. Gregg (Ed.), *Knowledge and cognition* (pp. 43–103). Oxford, UK: Lawrence Erlbaum.

Fang, Z. (1996). A review of research on teacher beliefs and practices. *Educational Research*, *38*(1), 47–65. doi:10.1080/0013188960380104

Fong, G. T., Krantz, D. H., & Nisbett, R. E. (1986). The effects of statistical training on thinking about everyday problems. *Cognitive Psychology*, *18*(3), 253–292. doi:10.1016/0010-0285(86)90001-0

Forbes, S., Camden, M., Pihama, N., Bucknall, P., & Pfannkuch, M. (2011). Official Statistics and statistical literacy: They need each other. *Statistical Journal of the International Association for Official Statistics*, *27*, 113–128.

Fox, J. J. (1860). On the province of the statistician. *Journal of the Statistical Society of London, 23*(3), 330–336. doi:10.2307/2338500

Gal, I. (Ed.). (2000). *Adult numeracy development: Theory, research, practice.* Cresskill, NJ: Hampton Press.

Gal, I. (2002). Adults statistical literacy: Meanings, components, responsibilities. *International Statistical Review, 70*(1), 1–25. doi:10.1111/j.1751-5823.2002.tb00336.x

Gal, I. (2003). Expanding conceptions of statistical literacy: An analysis of products from statistics agencies. *Statistics Education Research Journal, 2*, 3–21.

Gal, I., & Garfield, J. B. (Eds.). (1997). *The assessment challenge in statistics education* (Vol. 12). Amsterdam, Netherlands: IOS Press.

Garfield, J. B. (1993). Teaching statistics using small-group cooperative learning. *Journal of Statistics Education, 1*, 1–9.

Garfield, J.B. (1995). How students learn statistics. *International Statistical Review/Revue Internationale de Statistique, 63*, 25-34.

Garfield, J. B. (2002). The challenge of developing statistical reasoning. *Journal of Statistics Education, 10*, 58–69.

Garfield, J. B. (2003). Assessing statistical reasoning. *Statistics Education Research Journal, 2*, 22–38.

Garfield, J. B., & Ahlgren, A. (1988). Difficulties in learning basic concepts in probability and statistics: Implications for research. *Journal for Research in Mathematics Education, 19*(1), 44–63. doi:10.2307/749110

Garfield, J. B., & Ben-Zvi, D. (2007). How students learn statistics revisited: A current review of research on teaching and learning statistics. *International Statistical Review, 75*(3), 372–396. doi:10.1111/j.1751-5823.2007.00029.x

Garfield, J. B., & Ben-Zvi, D. (2009). Helping students develop statistical reasoning: Implementing a statistical reasoning learning environment. *Teaching Statistics, 31*(3), 72–77. doi:10.1111/j.1467-9639.2009.00363.x

Garfield, J. B., & Gal, I. (1999). Assessment and statistics education: Current challenges and directions. *International Statistical Review, 67*(1), 1–12. doi:10.1111/j.1751-5823.1999.tb00377.x

Garfield, J. B., Hogg, B., Schau, C., & Whittinghill, D. (2002). First courses in statistical science: The status of educational reform efforts. *Journal of Statistics Education, 10*, 456–467. Retrieved from https://ww2.amstat.org/publications/jse/v10n2/garfield.html

Gick, M. L., & Holyoak, K. J. (1980). Analogical problem solving. *Cognitive Psychology, 12*(3), 306–355. doi:10.1016/0010-0285(80)90013-4

Gigerenzer, G. (2008). Why heuristics work. *Perspectives on Psychological Science, 3*(1), 20–29. doi:10.1111/j.1745-6916.2008.00058.x PMID:26158666

Gigerenzer, G., & Hug, K. (1992). Domain-specific reasoning: Social contracts, cheating, and perspective change. *Cognition, 43*(2), 127–171. doi:10.1016/0010-0277(92)90060-U PMID:1617917

Goel, V., & Grafman, J. (1995). Are the frontal lobes implicated in planning functions? Interpreting data from the Tower of Hanoi. *Neuropsychologia, 33*(5), 623–642. doi:10.1016/0028-3932(95)90866-P PMID:7637857

Gronlund, N. E. (1998). *Assessment of student achievement.* Needham Heights, MA: Allyn & Bacon Publishing.

Groth, R. E. (2006). An exploration of students statistical thinking. *Teaching Statistics, 28*(1), 17–21. doi:10.1111/j.1467-9639.2006.00003.x

Hammerman, J. K., & Rubin, A. (2004). Strategies for managing statistical complexity with new software tools. *Statistics Education Research Journal, 3*, 17–41.

Harasim, L. (1996). Online education. In T. M., Harrison & T. Stephen (Eds.), *Computer networking and scholarly communication in the twenty-first-century university* (pp. 203- 214). Albany, NY: State University of New York.

Harrow, A. J. (1972). *A taxonomy of the psychomotor domain: A guide for developing behavioral objectives.* Boston, MA: Addison-Wesley Longman Ltd.

Haskell, R. E. (2000). *Transfer of learning: Cognition and instruction*. Cambridge, MA: Academic Press.

Hayes, J. R., & Simon, H. A. (1974). Understanding written problem instructions. In L. W. Gregg (Ed.), *Knowledge and cognition* (pp. 167–200). Oxford, UK: Lawrence Erlbaum Associates.

Heeger, D. J., & Ress, D. (2002). What does fMRI tell us about neuronal activity?. *Nature Reviews. Neuroscience*, *3*(2), 142–151. doi:10.1038/nrn730 PMID:11836522

Hinz, A. M. (1992). Shortest paths between regular states of the Tower of Hanoi. *Information Sciences*, *63*(1-2), 173–181. doi:10.1016/0020-0255(92)90067-I

Hinz, A. M., Klavžar, S., Milutinović, U., & Petr, C. (2013). *The tower of Hanoi–myths and maths*. Berlin, Germany: Springer Science & Business Media. doi:10.1007/978-3-0348-0237-6

Hoc, J. M. (1988). *Cognitive psychology of planning*. San Diego, CA: Academic Press Professional, Inc.

Hogg, R. V. (1991). Statistical education: Improvements are badly needed. *The American Statistician*, *45*, 342–343. doi:10.1080/00031305.1991.10475832

Hoskins, S. L., & Van Hooff, J. C. (2005). Motivation and ability: Which students use online learning and what influence does it have on their achievement?. *British Journal of Educational Technology*, *36*(2), 177–192. doi:10.1111/j.1467-8535.2005.00451.x

Hubbard, R. (1997). Assessment and the process of learning statistics. *Journal of Statistics Education*, *5*, 1–8.

Huck, S. W. (2007). Reform in statistical education. *Psychology in the Schools*, *44*(5), 527–533. doi:10.1002/pits.20244

Humes, G. E., Welsh, M. C., Retzlaff, P., & Cookson, N. (1997). Towers of Hanoi and London: Reliability and validity of two executive function tasks. *Assessment*, *4*(3), 249–257. doi:10.1177/107319119700400305 PMID:26613774

Iran-Nejad, A., McKeachie, W. J., & Berliner, D. C. (1990). The multisource nature of learning: An introduction. *Review of Educational Research*, *60*(4), 509–515. doi:10.3102/00346543060004509

Johnson, R. B., Onwuegbuzie, A. J., & Turner, L. A. (2007). Toward a definition of mixed methods research. *Journal of Mixed Methods Research, 1*(2), 112–133. doi:10.1177/1558689806298224

Jones, G. A., Langrall, C. W., Mooney, E. S., & Thornton, C. A. (2004). Models of development in statistical reasoning. In D. Ben-Zvi & J. B. Garfield (Eds.) *The challenge of developing statistical literacy, reasoning and thinking* (pp. 97-117). Dordrecht, Netherlands: Kluwer Academic Publishers.

Jones, G. A., Thornton, C. A., Langrall, C. W., Mooney, E. S., Perry, B., & Putt, I. J. (2000). A framework for characterizing childrens statistical thinking. *Mathematical Thinking and Learning, 2*(4), 269–307. doi:10.1207/S15327833MTL0204_3

Jueptner, M., & Weiller, C. (1995). Review: Does measurement of regional cerebral blood flow reflect synaptic activity?—Implications for PET and fMRI. *NeuroImage, 2*(2), 148–156. doi:10.1006/nimg.1995.1017 PMID:9343597

Jurado, M. B., & Rosselli, M. (2007). The elusive nature of executive functions: A review of our current understanding. *Neuropsychology Review, 17*(3), 213–233. doi:10.1007/s11065-007-9040-z PMID:17786559

Kahneman, D., Slovic P., & Tversky. (1982). *Judgment under uncertainty: Heuristics and biases*. Cambridge, UK: Cambridge University Press.

Kahneman, D., & Tversky, A. (1982). On the study of statistical intuitions. *Cognition, 11*(2), 123–141. doi:10.1016/0010-0277(82)90022-1 PMID:7198957

Karppinen, P. (2005). Meaningful learning with digital and online videos: Theoretical perspectives. *Association for the Advancement of Computing in Education Journal, 13*, 233–250.

Keeler, C. M., & Steinhorst, R. K. (1995). Using small groups to promote active learning in the introductory statistics course: A report from the field. *Journal of Statistics Education, 3*, 1–8. Retrieved from http://ww2.amstat.org/publications/jse/v3n2/Keeler.html

Knauff, M., Fangmeier, T., Ruff, C. C., & Johnson-Laird, P. N. (2003). Reasoning, models, and images: Behavioral measures and cortical activity. *Journal of Cognitive Neuroscience, 15*(4), 559–573. doi:10.1162/089892903321662949 PMID:12803967

Kotovsky, K., Hayes, J. R., & Simon, H. A. (1985). Why are some problems hard? Evidence from Tower of Hanoi. *Cognitive Psychology, 17*(2), 248–294. doi:10.1016/0010-0285(85)90009-X

Krampe, R. T. (2002). Aging, expertise and fine motor movement. *Neuroscience and Biobehavioral Reviews, 26*(7), 769–776. doi:10.1016/S0149-7634(02)00064-7 PMID:12470688

Krathwohl, D. R. (2002). A revision of Blooms taxonomy: An overview. *Theory into Practice, 41*(4), 212–218. doi:10.1207/s15430421tip4104_2

Krathwohl, D. R., Bloom, B. S., & Masia, B. B. (1964). *Taxonomy of educational objectives, handbook ii: affective domain*. New York, NY: David McKay Company.

Kulm, G. (1990). *Assessing higher order thinking in mathematics*. Waldorf, MD: American Association for the Advancement of Science Books.

Kunda, Z., & Nisbett, R. E. (1986). Prediction and the partial understanding of the law of large numbers. *Journal of Experimental Social Psychology, 22*(4), 339–354. doi:10.1016/0022-1031(86)90019-3

Lambos, C., & Delfabbro, P. (2007). Numerical reasoning ability and irrational beliefs in problem gambling. *International Gambling Studies, 7*(2), 157–171. doi:10.1080/14459790701387428

Larwin, K., & Larwin, D. (2011). A meta-analysis examining the impact of computer-assisted instruction on postsecondary statistics education: 40 years of research. *Journal of Research on Technology in Education, 43*(3), 253–278. doi:10.1080/15391523.2011.10782572

Link, T. M., & Marz, R. (2006). Computer literacy and attitudes towards e-learning among first year medical students. *BMC Medical Education, 6*(1), 34. doi:10.1186/1472-6920-6-34 PMID:16784524

Lovie, P., & Lovie, A. D. (1976). Teaching intuitive statistics I: Estimating means and variances. *International Journal of Mathematical Education in Science and Technology, 7*(1), 29–39. doi:10.1080/0020739760070104

Mahmoudi, A., Takerkart, S., Regragui, F., Boussaoud, D., & Brovelli, A. (2012). Multivoxel pattern analysis for FMRI data: A review. *Computational and Mathematical Methods in Medicine, 2012*, 1–14. doi:10.1155/2012/961257 PMID:23401720

Makrymichalos, M., Antony, J., Antony, F., & Kumar, M. (2005). Statistical thinking and its role for industrial engineers and managers in the 21st century. *Managerial Auditing Journal*, *20*(4), 354–363. doi:10.1108/02686900510592043

Marcal, L., & Roberts, W. W. (2000). Computer literacy requirements and student performance in business communications. *Journal of Education for Business*, *75*(5), 253–257. doi:10.1080/08832320009599023

Marzocchi, A. S. (2016). Using the Tower of Hanoi puzzle to infuse your mathematics classroom with computer science concepts. *International Journal of Mathematical Education in Science and Technology*, *47*(5), 814–821. do i:10.1080/0020739X.2015.1133855

Matthews, P. M., & Jezzard, P. (2004). Functional magnetic resonance imaging. *Journal of Neurology, Neurosurgery, and Psychiatry*, *75*, 6–12. PMID:14707297

Mayer, R. E. (2004). Should there be a three-strikes rule against pure discovery learning?. *The American Psychologist*, *59*(1), 14–19. doi:10.1037/0003-066X.59.1.14 PMID:14736316

Mevarech, Z. R. (1983). A deep structure model of students statistical misconceptions. *Educational Studies in Mathematics*, *14*(4), 415–429. doi:10.1007/BF00368237

Mills, J. D., & Raju, D. (2011). Teaching statistics online: A decade's review of the literature about what works. *Journal of Statistics Education*, *19*, 1–28. Retrieved from http://ww2.amstat.org/publications/jse/v19n2/mills.pdf

Mills, J. D., & Xu, Y. (2006). Statistics at a distance: Technological tools, learning, and design features for todays modern course. *Journal of Educational Technology Systems*, *34*(4), 427–446. doi:10.2190/DHKE-8AEC-76GA-XBAT

Moore, D. S. (1997). New pedagogy and new content: The case of statistics. *International Statistical Review*, *65*(2), 123–137. doi:10.1111/j.1751-5823.1997.tb00390.x

Moore, D. S., Cobb, G. W., Garfield, J., & Meeker, W. Q. (1995). Statistics education fin de siècle. *The American Statistician*, *49*, 250–260.

Moore, T. (Ed.). (2001). *Teaching statistics: Resources for undergraduate instructors*. Washington, DC: Mathematical Association of America and American Statistical Association.

Newell, A., & Simon, H. A. (1972). *Human problem solving* (Vol. 104). Englewood Cliffs, NJ: Prentice-hall.

Newman, S. D., Twieg, D. B., & Carpenter, P. A. (2001). Baseline conditions and subtractive logic in neuroimaging. *Human Brain Mapping*, *14*(4), 228–235. doi:10.1002/hbm.1055 PMID:11668654

Nisbett, R. E., Krantz, D. H., Jepson, C., & Kunda, Z. (1983). The use of statistical heuristics in everyday inductive reasoning. *Psychological Review*, *90*(4), 339–363. doi:10.1037/0033-295X.90.4.339

Norman, K. A., Polyn, S. M., Detre, G. J., & Haxby, J. V. (2006). Beyond mind-reading: Multi- voxel pattern analysis of fMRI data. *Trends in Cognitive Sciences*, *10*(9), 424–430. doi:10.1016/j.tics.2006.07.005 PMID:16899397

Oakes, M. (1986). *Statistical inference: A commentary for the social and behavioural sciences*. Chichester, UK: Wiley.

Owen, A. M. (1997). Cognitive planning in humans: Neuropsychological, neuroanatomical and neuropharmacological perspectives. *Progress in Neurobiology*, *53*(4), 431–450. doi:10.1016/S0301-0082(97)00042-7 PMID:9421831

Park, D. C., & Reuter-Lorenz, P. (2009). The adaptive brain: Aging and neurocognitive scaffolding. *Annual Review of Psychology*, *60*(1), 173–196. doi:10.1146/annurev.psych.59.103006.093656 PMID:19035823

Perkins, D. N., & Salomon, G. (1992). Transfer of learning. In T. N. Postlethwaite & T. Husen (Eds.), *International encyclopedia of education* (Vol. 2, pp. 6452–6457). Oxford, UK: Pergamon Press.

Peterson, C. R., & Beach, L. R. (1967). Man as an intuitive statistician. *Psychological Bulletin*, *68*(1), 29–46. doi:10.1037/h0024722 PMID:6046307

Pfannkuch, M., & Rubick, A. (2002). An exploration of students' statistical thinking with given data. *Statistics Education Research Journal*, *1*, 4–21.

Pfannkuch, M., & Wild, C. (2004). Towards an understanding of statistical thinking. In D. Ben-Zvi & J. B. Garfield (Eds.) *The challenge of developing statistical literacy, reasoning and thinking* (pp. 17-46). Dordrecht, Netherlands: Kluwer Academic Publishers. doi:10.1007/1-4020-2278-6_2

Pfannkuch, M., & Wild, C. J. (2000). Statistical thinking and statistical practice: Themes gleaned from professional statisticians. *Statistical Science*, *15*, 132–152.

Ploger, D., & Wilson, M. (1991). Statistical reasoning: What is the role of inferential rule training? Comment on Fong and Nisbett. *Journal of Experimental Psychology. General*, *120*(2), 213–214. doi:10.1037/0096-3445.120.2.213

Poldrack, R. A. (2008). The role of fMRI in cognitive neuroscience: Where do we stand?. *Current Opinion in Neurobiology*, *18*(2), 223–227. doi:10.1016/j.conb.2008.07.006 PMID:18678252

(2004). Research on statistical literacy, reasoning, and thinking: Issues, challenges, and implications. InGarfield, J. B., & Ben-Zvi, D. (Eds.), *The challenge of developing statistical literacy, reasoning and thinking* (pp. 397–409). Dordrecht, Netherlands: Kluwer Academic Publishers.

Resnick, L. B., & Resnick, D. P. (1992). Assessing the thinking curriculum: New tools for educational reform. In B. Gilford & M.C. O'Conner (Eds.), *Changing assessments* (pp. 37-75). Springer.

Richardson, J., & Swan, K. (2003). Examing social presence in online courses in relation to students' perceived learning and satisfaction. *Journal of Asynchronous Learning Networks*, *7*, 68–88.

Robinson, S. J., & Brewer, G. (2016). Performance on the traditional and the touch screen, tablet versions of the Corsi Block and the Tower of Hanoi tasks. *Computers in Human Behavior*, *60*, 29–34. doi:10.1016/j.chb.2016.02.047

Rogoff, B. (1998). *Cognition as a collaborative process.* In W. Damon (Ed.), Handbook of child psychology (Vol. 2, pp. 679–743). New York, NY: Wiley.

Rumsey, D. J. (2002). Statistical literacy as a goal for introductory statistics courses. *Journal of Statistics Education*, *10*, 6–13.

Sandoval, W. A., & Bell, P. (2004). Design-based research methods for studying learning in context: Introduction. *Educational Psychologist*, *39*(4), 199–201. doi:10.1207/s15326985ep3904_1

Schmidt, F. L. (1996). Statistical significance testing and cumulative knowledge in psychology: Implications for training of researchers. *Psychological Methods*, *1*(2), 115–129. doi:10.1037/1082-989X.1.2.115

Schmidt, R. A., & Lee, T. (1988). *Motor control and learning* (5th ed.). Champaign, IL: Human kinetics.

Scholnick, E. K., Friedman, S. L., & Wallner-Allen, K. E. (1997). What do they really measure? A comparative analysis of planning tasks. In S. L. Friedman & E. K. Scholnick (Eds.), *The developmental psychology of planning: Why, how, and when do we plan* (pp. 127-156). New York, NY: Psychology Press.

Seddon, G. M. (1978). The properties of Blooms taxonomy of educational objectives for the cognitive domain. *Review of Educational Research*, *48*(2), 303–323. doi:10.3102/00346543048002303

Shale, D. (1990). Toward a reconceptualization of distance education. In M. G. Moore (Ed.), *Contemporary issues in American distance education* (pp. 333–343). Oxford, UK: Pergamon Press.

Silver, N. (2012). *The signal and the noise: Why so many predictions fail-but some don't*. New York, NY: Penguin.

Simon, H. A. (1975). The functional equivalence of problem solving skills. *Cognitive Psychology*, *7*(2), 268–288. doi:10.1016/0010-0285(75)90012-2

Simon, H. A. (1978). Information-processing theory of human problem solving. In W. K. Estes (Ed.), *Handbook of learning and cognitive processes* (Vol. 5, pp. 271–295). Mahwah, NJ: Lawrence Erlbaum Associates.

Simpson, E. (1972). *The psychomotor domain*. Washington, DC: Gryphon House.

Skinner, B. F. (1987). Whatever happened to psychology as the science of behavior? *The American Psychologist*, *42*(8), 780–786. doi:10.1037/0003-066X.42.8.780

Smart, K. L., & Cappel, J. J. (2006). Students' perceptions of online learning: A comparative study. *Journal of Information Technology Education*, *5*, 201–219.

Song, L., Singleton, E. S., Hill, J. R., & Koh, M. H. (2004). Improving online learning: Student perceptions of useful and challenging characteristics. *The Internet and Higher Education*, *7*(1), 59–70. doi:10.1016/j.iheduc.2003.11.003

Stark, C. E., & Squire, L. R. (2001). When zero is not zero: The problem of ambiguous baseline conditions in fMRI. *Proceedings of the National Academy of Sciences of the United States of America*, *98*(22), 12760–12766. doi:10.1073/pnas.221462998 PMID:11592989

Stroud, J. B. (1940). Experiments on learning in school situations. *Psychological Bulletin*, *37*(10), 777–807. doi:10.1037/h0063014

Swan, K. (2001). Virtual interaction: Design factors affecting student satisfaction and perceived learning in asynchronous online courses. *Distance Education*, *22*(2), 306–331. doi:10.1080/0158791010220208

Sweller, J. (1988). Cognitive load during problem solving: Effects on learning. *Cognitive Science*, *12*(2), 257–285. doi:10.1207/s15516709cog1202_4

Tishkovskaya, S., & Lancaster, G. A. (2012). Statistical education in the 21st century: A review of challenges, teaching innovations and strategies for reform. *Journal of Statistics Education*, *20*, 1–56.

Tong, F., & Pratte, M. S. (2012). Decoding patterns of human brain activity. *Annual Review of Psychology*, *63*(1), 483–509. doi:10.1146/annurev-psych-120710-100412 PMID:21943172

Von Roten, F. C. (2006). Do we need a public understanding of statistics?. *Public Understanding of Science (Bristol, England)*, *15*(2), 243–249. doi:10.1177/0963662506061883

Vul, E., Harris, C., Winkielman, P., & Pashler, H. (2009). Puzzlingly high correlations in fMRI studies of emotion, personality, and social cognition. *Perspectives on Psychological Science*, *4*(3), 274–290. doi:10.1111/j.1745-6924.2009.01125.x PMID:26158964

Wallman, K. K. (1993). Enhancing statistical literacy: Enriching our society. *Journal of the American Statistical Association*, *88*, 1–8. doi:10.1080/01621459.1993.10594283

Warner, C. B., & Meehan, A. M. (2001). Microsoft Excel™ as a tool for teaching basic statistics. *Teaching of Psychology*, *28*(4), 295–298. doi:10.1207/S15328023TOP2804_11

Watkins, A. E., Scheaffer, R. L., & Cobb, G. W. (2010). *Statistics: From data to decision*. Hoboken, NJ: John Wiley & Sons.

Watson, J., & Callingham, R. (2003). Statistical literacy: A complex hierarchical construct. *Statistics Education Research Journal*, *2*, 3–46.

Watson, J. M. (1997). Assessing statistical thinking using the media. In I. Gal & J. B. Garfield (Eds.), *The assessment challenge in statistics education* (Vol. 12, pp. 107–121). Amsterdam, Netherlands: IOS Press.

Welsh, M. C., Satterlee-Cartmell, T., & Stine, M. (1999). Towers of Hanoi and London: Contribution of working memory and inhibition to performance. *Brain and Cognition*, *41*(2), 231–242. doi:10.1006/brcg.1999.1123 PMID:10590821

Weltman, D., & Whiteside, M. (2010). Comparing the effectiveness of traditional and active learning methods in business statistics: Convergence to the mean. *Journal of Statistics Education*, *18*, 1–13.

Wild, C. J., & Pfannkuch, M. (1999). Statistical thinking in empirical enquiry. *International Statistical Review*, *67*(3), 223–248. doi:10.1111/j.1751-5823.1999.tb00442.x

Wood, D., Bruner, J. S., & Ross, G. (1976). The role of tutoring in problem solving. *Journal of Child Psychology and Psychiatry, and Allied Disciplines*, *17*(2), 89–100. doi:10.1111/j.1469-7610.1976.tb00381.x PMID:932126

Wood, E., Willoughby, T., Rushing, A., Bechtel, L., & Gilbert, J. (2005). Use of computer input devices by older adults. *Journal of Applied Gerontology*, *24*(5), 419–438. doi:10.1177/0733464805278378

Woodworth, R. S., & Thorndike, E. L. (1901). The influence of improvement in one mental function upon the efficiency of other functions (I). *Psychological Review*, *8*(3), 247–261. doi:10.1037/h0074898

Xu, Y., & Corkin, S. (2001). HM revisits the Tower of Hanoi Puzzle. *Neuropsychology*, *15*(1), 69–79. doi:10.1037/0894-4105.15.1.69 PMID:11216891

Zhao, Y., Lei, J., Yan, B., Lai, C., & Tan, H. S. (2005). What makes the difference? A practical analysis of research on the effectiveness of distance education. *Teachers College Record*, *107*(8), 1836–1884. doi:10.1111/j.1467-9620.2005.00544.x

Zook, N. A., Davalos, D. B., DeLosh, E. L., & Davis, H. P. (2004). Working memory, inhibition, and fluid intelligence as predictors of performance on Tower of Hanoi and London tasks. *Brain and Cognition*, *56*(3), 286–292. doi:10.1016/j.bandc.2004.07.003 PMID:15522766

ENDNOTE

1 Consider the probability of cheating where C means cheated; p(Dishonest|C). We know p(dishonest) = 0.001, p(honest) = 0.999, p(C|dishonest) = 1.00, p(C|honest) = 0.05. Applying *Bayes Theorem* (see Silver, 2012), the probability that this student cheated was 0.0196, or 2% (with a prior probability of 1.00 * 0.001 + 0.05 * 0.999 = 0.051)

Chapter 3
Developing Statistics Cognitions

ABSTRACT

This chapter focuses on the development of statistics cognition, framing the discussion on the need to enhance cognitive development. We argue that by understanding how cognitive processes have been shown to inhibit learning, we can differentiate between types of errors in statistics education. By understanding the operationalized difference of a bias and misconception, those interested in statistics education can identify the sources of these errors, and subsequently develop a means to attenuate their effect. Using dual process theory, we argue that classifying the source of errors and differentiating between biases and misconceptions educators can use errors to enhance the development of statistics literacy, reasoning, and thinking.

Although none of us can know the future, predicting it is so important that organisms are hard wired to learn about predictable events in their environment. - Andy Field, 2013, p. 197

1. INTRODUCTION

If the purpose of statistics education is to improve statistics cognitions, then it is imperative to have a detailed and specific model, which makes use of interdisciplinary work to connect cognitive features to educational outcomes.

DOI: 10.4018/978-1-5225-2420-5.ch003

With a model in mind (Chapter 2), we can now begin to understand how we can develop statistics cognitions. When we refer to "*developing*" statistics cognitions, we are talking about enhancing cognitive processes. This differs from classes framed on traditional behavioral curriculum, which arguably has focused more on learning content than learning skills (Moore, 1997). While this propensity to use content learning as a framework for statistics may have emerged due to ease, the development of cognitive processes are of critical importance to the future of a statistics education framed on cognitive objectives.

Thus, to understand how educators can develop statistics cognitions, this Chapter first addresses how cognitions can be adjusted through learning (2). Throughout this section, our discussion centers on the delineation among biases, misconceptions, and their sources (2.1), and uses dual process theory (2.2) to explain how errors could be useful for the field of statistics education (2.3).

We then turn to a more formal discussion on educating statistics cognitions, with emphasis placed on how cognitions are developed over time (3.1), and considerations needed to develop cognitions in an online learning environment (3.2). We then move on to discussion how statistics cognitions can be enhanced in online environments (4), ending with a summary of this Chapter (5).

2. COGNITIONS USED TO LEARN

The ultimate goal of statistics education has become to enhance statistics cognitions. As discussed, quantifying cognitions is difficult (Chapter 2), but is of central interest to educators and researchers who have a vested interested in statistics education. Although ways that cognition could be quantified have been outlined, an important assumption made by a cognitively derived curriculum is that cognitions can be enhanced. Namely, students in a statistics class should retain working knowledge of important statistical concepts, be able to demonstrate sound rationale for their statistical practices, and be able to apply a set of skills relevant toward statistics by the end of the course (Kundert, 1990; Mallows, 1998; Woolley, 2013). However, students come into classrooms with their own cognitions, often related to factors that an educator has no reasonable control over (Wright et al., 2001). Thus, it is not enough to establish the types of cognitions we would like to see from students, but understand how cognitions themselves are part of the learning process.

2.1. How Cognitions Can Harm Learning

With an increasingly diverse population of students taking a statistics course (Horn, Peter, & Rooney, 2002; Perkins & Saris, 2001), many scholars have looked for ways to improve learning statistics (Garfield, 1995; Li & Shen, 1992; Magel, 1996). Although students often express difficulty in learning statistics (Garfield & Ahlgren, 1988) significant work has focused on how such difficulties can impair learning statistics (Garfield & Ben-Zvi, 2007; Lovett & Greenhouse, 2000; Snee, 1993). Yet, these difficulties need not only be in content, but may arise from the way people implicitly think about statistics (see Tversky, Kahneman, & Moser, 1990).

For example, consider he gambler's fallacy (Ayton & Fischer, 2004; Lehrer, 2009), a cognitive bias whereby individuals think something is more (or less) likely to occur, simply because of the events that preceded it. Although it is easy to consider how such an effect could be demonstrated in a laboratory, such an effect may even be seen in recent worldly events.

In the Feburary of 2016, the Democratic National Committee was preoccupied dealing with an internal crisis, namely disputes among those who supported the their parties two nominees; Former Secretary of State and Senator Hillary Clinton, and Senator Bernie Sanders. In the hotly contested Iowa caucus for the Democratic Presidential Nomination, there was a little known (and far less practiced) rule that in a "deadlocked" caucus, where participants did not agree on a single candidate, would have their the delegates assigned based on the flip of a coin. When it was reported that Clinton won 6 out of 6 tosses, Sanders supporters cried foul (Norton, 2016), arguing that Clinton had only a 1 in 6 chance of winning all 6 tosses. Arguably, as coin tosses happened, Sander's supporters may have felt that based on chance alone, he should have won 3 of the tosses. However, these coin tosses occurred by designated appointees, in different parts of the state, at different times, using different coins, with each candidate presumably being randomly assigned as to whether they were "*heads*" or "*tails*". While some Sander's proponents felt the coin tosses had been rigged in Clinton's favor, it may be more intuitive to assert that these coin tosses were not mutually inclusive, and that the outcome of one coin toss did not, in fact, influence the next.

Indeed, as illustrated by this example, a little statistical literacy, reasoning, and thinking, could have gone a long way. Although one could be tempted to claim such a finding may be unique to the demographic that supported Sanders, cognitions can create blockades whereby any person overlooks important statistical nuances (Sunstein, Bobadilla-Suarez, Lazzaro, & Sharot, 2016). For example, Sunstein and colleagues conducted a study that provided participants with climate change information, and asked a series of questions on their beliefs regarding climate change, and asked to estimate how much the climate will change by 2020. Not too surprisingly, Sunstein found that some participants strongly believed in climate change, predicting a larger increase in average temperature, and some strongly believed that man made climate change was a farce, predicting a low amount of temperature variation, and others were in the middle.

Interestingly, when presented with new information, that indicated that the previously stated problem of climate change was either greatly exaggerated, or much worse than previously stated (depending on condition), participants only used the newly presented information to update their predictions if it aligned with their previously stated beliefs. That is, those who strongly believed climate change would occur reported a larger degree of temperature change when they received news that the problem was worse, but did not lower their original predictions if the news stated previous findings had been exaggerated. Conversely, those who maintained strong beliefs that man made climate change was an artifact only used information that said that the previous findings had been exaggerated when updating their predictions, only lowering the amount of temperature change that they felt was likely to occur. Further, these results did not differ by political affiliation when demographic variables were taken into account.

These two examples illustrate two key points. First, they illustrate the dangers of ignoring the societal development of statistics cognitions. For centuries, scholars have recognized the importance of a statistically literate and rationale populace (Dewey, 1891). Second, these examples also suggest our cognitions can be unduly biased by a variety of factors, including abstract differences that exist at the individual level, such as personal beliefs. This can be highly problematic for educators, as such characteristics reside in the student, and are less likely to be altered by an educator, who is conceivably charged with overseeing the cognitive development of hundreds of students at a time. While it may seem apparent to study what goes right within the context of education, scholarly advances have often come about by studying the types of mistakes students make (Kunda & Nisbett, 1986). Indeed, one of the primary objectives of facilitating learning within statistics education is to understand how biases and misconceptions can arise throughout the learning process.

2.1.1. Cognitive Biases and Misconceptions

In order to adequately discuss bias and misconceptions, it is important to define both terms. Although previous scholars have not explicitly delineated between bias and misconception (Garfield, 2003; Hirsch & O'Donnell, 2001; Sotos, Vanhoof, Van den Noortgate, & Onghena, 2007), it seems reasonable that such a distinction is necessary.

Specifically, biases refer to illogical choices made about inferences (Haselton, Nettle, & Andrews, 2005). These errors are usually explored in terms of decision making (Kahneman & Tversky, 1996), and are thought to act as mental shortcuts enabling people to make faster decisions (Tversky & Kahneman, 1974). Conversely, misconceptions are erroneous beliefs, that are widely held by a population. Importantly, the source of the misconception is not specified by the presence of the misconception itself. Although a student may have a difficult time understanding the nuances of significance testing (Oakes, 1986) or how to read confidence intervals (Cumming, Williams, & Fidler, 2004), such errors could be derived from cognitive biases, a lack of knowledge, or even an erroneous belief about the meaning or application of knowledge or skill.

This distinction, although seemingly tautological in nature, has important ramifications for how scholars and educators can understand statistics cognition. In particular, if misconceptions are derived from incorrect content, whereas biases are derived from internal cognitive processes, then at the very least, educators should approach these errors differently (see 2.3). As a result, it is important to identify sources of various types of errors, so that we can then discuss how to attenuate their negative influence on learning.

2.1.2. Sources of Error

With an increasingly diverse population of students taking a statistics course (Horn et al., 2002; Perkins & Saris, 2001), many scholars have looked at ways to improve learning statistics (Garfield, 1995; Li & Shen, 1992; Magel, 1996). Yet because errors in understanding can happen in a variety of ways (Batanero, Godino, Vallecillos, Green, & Holmes, 1994), some scholars have sought to focus their research on how specific errors arise (Sotos et al., 2007). Yet biases and misconceptions are as diverse as the very students they take hold of[1], prompting many to disagree as to how biases can be understood and categorized (Dougherty, Gettys, & Ogden, 1999; Kahneman & Tversky,

1996), with considerably less work aimed at distinguishing biases from misconceptions.

Yet, by making this distinction, educators can help shed light on how certain errors arise and what they can do to reduce their negative effects. For instance, if a student responds incorrectly due to a cognitive bias, their incorrect response does not necessarily imply a lack of knowledge. Instead, the response could be due to a lack of cognitive resources to correct for a bias. Although we will momentarily discuss the ramifications of biases and misconceptions for educators, what's important is understanding that while cognitive biases are deviations from logical thought (Haselton et al., 2005; Kahneman & Tversky, 1982), people are often not logical in their thinking (see Kahneman, 2011). As such, an incorrect response derived from a cognitive bias is not necessarily a sign of a content educational deficiency, but rather the failure of a monitoring process.

While such a failure is alarming, the mechanisms behind cognitive biases could have evolved with an adaptive purpose (Haselton et al., 2005). Some scholars have claimed that such biases are the result of mental shortcuts (Simon, Houghton, & Aquino, 2000) that were formed to expedite the decision making process (Tversky & Kahneman, 1974). This line of thinking is parallel with other sub-disciplines within psychological research, for scholars have proposed that people are "*cognitive misers*" (Fiske & Taylor, 1991) who may choose to sacrifice accuracy (Tversky & Kahneman, 1973a; 1974) in order to navigate the social world with greater efficiency.

Conversely, unlike cognitive biases, misconceptions have less to do with *attributions* or *heuristics (*Kahneman & Tversky, 1996) and more to do with misinterpretations (Gardenier & Resnik, 2002). Importantly, these misinterpretations may or may not be the result of cognitive biases. To illustrate the difference between a bias and misconception consider the dual process cognitive framework (Morewedge & Kahneman, 2010), asserting that while cognitive biases may come from any combination of heuristics (Kahneman, Slovic, & Tversky, 1982), limited processing capacity (Simon, 1955), emotional motivations (Pfister & Böhm, 2008), and social forces (Wang, Simmons, & Brédart, 2001), they can also arise from a limited capacity to think. This *dual-system model of reasoning*, attempts to elucidate biases in terms of automated and controlled systems, that can arguably be applied to the monitoring of misconceptions.

2.2. Understanding the Influence of Errors: Dual Process Theory

The dual model process of reasoning implies that automated systems constantly generate responses that the controlled system can then monitor, modify, reject, or approve. Controlled systems are used to think carefully about tasks, but automated systems are designed to be acted upon (Kahneman & Klein, 2009), offering a high-quality, readily available judgment, when time is of the essence (Dijksterhuis, Bos, Nordgen, & Van Baaren, 2006; Dijksterhuis & Meurs, 2006; Dijksterhuis & Nordgren, 2006).

Although automated processes may often provide excellent responses, scholars have examined reoccurring patterns of when they do not (Kahneman et al., 1982; Tversky & Kahneman 1973b). Thus, questions regarding a failure of controlled processes have been examined by researchers attempting to uncover why cognitive biases are so pervasive to human decision making (Baumeister & Heatheron, 1996). However, a failure to use controlled processes is not the only type of error that can occur. As depicted in a theorized model of these processes (Figure 1), problems also arise when controlled processes are

Figure 1. Theorized model of statistical cognitive biases and misconceptions

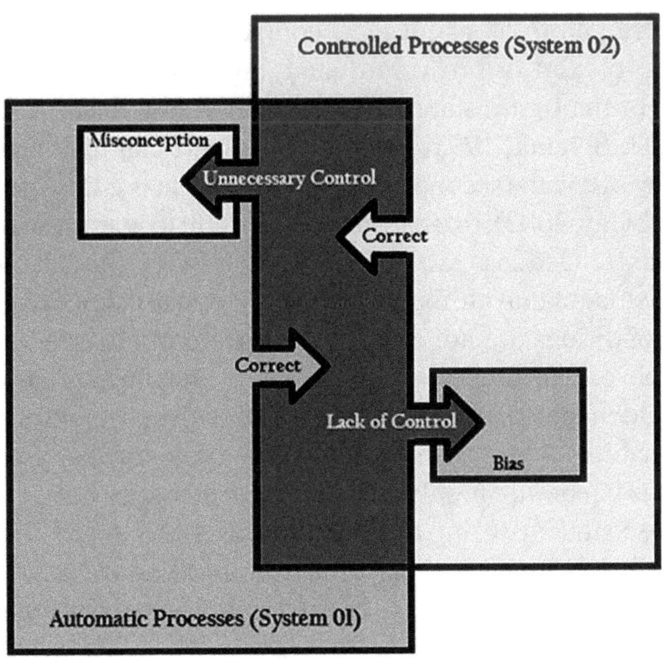

unnecessarily active. We maintain that this difference in the activation of the controlled processes not only defines the difference between misconceptions and cognitive biases, but that it has implications for specific ways educators counteract their negative effects, while capitalizing on their benefits.

As discussed, this framework asserts that cognitive biases are automated responses, designed to allow one to think fast (Kahneman, 2011), so that one can act quickly (Kahneman & Klein, 2009) without having to dedicate many resources. However, sometimes the responses generated are not adequate (Piattelli-Palmarini, 1994; Tversky & Kahneman, 1974), and could lead to at least two forms of error.

2.2.1. Classifications of Errors

One type of error that could occur is when an automated process generates a response, and the controlled system does not correct or prevent the response. This type of error is what can be thought of as a cognitive bias, because the automated response was incorrect, but the student failed to use a controlled processes, resulting in an error.

There could be several types of reasons for such an error, but the rationale behind the type of error is critical for an educator to understand, and is what educators should focus on to help students succeed. For instance, compare how an educator may help a student who is not motivated to approach their work with care and think about their coursework critically, relative to one who chooses to exert effort toward studying in an environment rich with distractions. In the first instance, student motivation needs to be addressed (Pajares, 2003; Schunk, 1991), such that students can appreciate how their efforts can pay off. In the second instance, the student is guilty of multi-tasking (Hembrook & Gay, 2003), that can make it difficult to learn content (Bowman, Levine, Waite, & Gendron, 2010; Strayer, Drews, & Crouch, 2006). In this instance, the student may think they are studying hard, but fail to realize that their study conditions are not optimal, and perform less well than they feel they should have. This issue has less to do with an internal characteristic that modulates student performance, but has to do with self-control and adopting good study habits. While both students may be more likely to rely on gut reactions, thus becoming highly susceptible to statistics biases, the educator may help these students using different approaches.

Additionally, errors are not just a matter of failing to use of a controlled process. As depicted in Figure 1, errors could also occur when controlled processes are overactive. For example, a growing body of research has suggested that controlled processes are similar to that of a muscle (Muraven & Baumeister, 2000), and may disrupt or inhibit one's ability to use processes such as self-control (Baumeister, Vohs, Tice, 2007; Gailliot & Baumeister, 2007; Hagger, Wood, Stiff, & Chatzisarantis, 2010; Muraven, Shmueli, Burkley, 2006). Although this may initially appear to be a cognitive bias in that the automated response is not corrected or modified by the controlled response, the actual reason for the error is that the controlled processes were used so frequently that they had become exhausted, rendering the student unable to use their controlled thought. Again, such an error could come from many sources, such as a failure to use self-regulatory strategies efficiently, however, this example illustrates the importance of understanding how students apply self-regulatory strategies, and what effect these strategies have on their overall learning.

2.2.2. Potential Mechanisms of Errors

There are several ways errors could be generated; however, there are numerous reasons for how these automated and controlled systems can break down. Whether by disability (Harris & Graham, 1985), a lack of motivation (Schunk, 1991), or exhaustion (Muraven & Baumeister, 2000), controlled processes can easily facilitate educational errors. By describing how errors occur, we maintain that scholars and educators can better understand their impact on education, and consider ways to prevent their negative influence on educational outcomes.

For instance, consider a situation where a student chooses not to critically think about the content they are learning. If the course is designed on cognitive objectives, a refusal to engage in the material could have a negative impact on that student's learning. In this situation, a controlled process, such as critically assessing the information presented on an exam, could be under active. Because this relative lack of monitoring a controlled process is related

Table 1. Statistical misconception and bias error, source, and field of impact

	Error	**Deficiency Source**	**Field Impact**
Bias	Under active (Choice)	Effort	Motivation
	Under active (No Choice)	Resource depletion	Self-regulation
Misconception	Over active (Choice)	Load / Bottlenecking	Anxiety

to the choices the student made, educators can center their efforts on enhancing student motivation (Lavasani, Weisani, & Shariati, 2014; Snee, 1993). While there are a variety of ways one can assess and improve student motivation (Chapter 6), online statistics educators must be careful in considering how the structure and design of their course can be used to motivate students (Bolliger, Supanakorn, & Boggs, 2010; Hodges, 2004; Wang & Reeves 2007).

Further, one can also imagine a situation that a student is unable to use a controlled strategy because they have depleted the mental energy needed to monitor and correct automatic responses (Muraven & Baumeister, 2000). A lack of resources could exist due to a student's individual cognitive ability, such as any condition that may negatively influence a student's working memory capacity (Passolunghi & Siegel, 2004; Wilson & Swanson, 2001). However, an inability to use a controlled process could also be due to exhaustion (Baddeley, 2003; Persson, Welsh, Jonides, & Reuter-Lorenz, 2007; Schmeichel, Vohs, & Baumeister, 2003). Regardless of the cause, an under active controlled process not due to a student's own choices, is likely to influence numerous outcomes of statistical education. For example, without the ability to use controlled processes, students could become more anxious (Carver & Scheier, 1988), as well as perform less well (Paas, Tuovinen, Tabbers, & Van Gerven, 2003).

Again, in this instance, educators may find themselves facing different educational challenges, based on the root cause. For instance, one reason a student could be exhausted is that they have inappropriately applied the incorrect study habits for the assessments of a course. For example, consider a student who makes flash cards when an exam will be comprised entirely of essays. The flash cards may have taken a while, and seemed like a lot of work to the student, but without making connections among these terms and critically thinking about what they mean, the student may not be as prepared as they feel. In such an instance, this student could benefit from

learning how to apply the appropriate self-regulation strategy to ensure their success (Brown, Roediger, & McDaniel, 2014; Dunlosky, Rawson, Marsh, Nathan, & Willingham, 2013; Karpicke, Butler, & Roediger, 2009; Weinstein, McDermott, & Roediger, 2010). Conversely, in a situation where a student has impairments with working memory, targeted interventions - such as mindfulness training - could help students succeed (Mrazek, Franklin, Phillips, Baird, & Schooler, 2013).

Although discussion has focused on under activity in controlled processes, it is conceivable that errors arise when controlled processes are over used. For instance students who concurrently use cognitive processes may find themselves at risk of diminished performance (Hembrook & Gay, 2003; Strayer et al., 2006; Strayer & Drews, 2007). By dividing cognitive resources (i.e., attentional control), students may find themselves failing to encode information correctly, and develop enduring misunderstandings of concepts that are necessary to further one's understanding of statistics. Such a process, referred to as *cognitive bottlenecking* (see Borst, Taatgen, & van Rijn, 2010), occurs when there is an inability to attend to specific information because of limited resources available while processing information. Such a finding is not only a concern for memorizing content, as studies have suggested that when attentional resources are limited, students may have more difficulty in thinking critically about the information they encode (Allport, 1993; Pashler, 1984; Posner & Snyder, 1974).

Assuming the structure of dual process theory, it seems clear that activity of controlled processes is of key importance to scholars and educators who want to improve learning. These errors potentially contain insights that can target remediation and help students learn more efficiently. Indeed, there are many ways errors can be used in statistics education.

2.3. The Use of Errors in Statistics Education

Because errors are based on cognitive biases and content misconceptions, there are likely different ways cognitions and misconceptions can influence one's ability to learn statistics. In order to develop an understanding of how biases and misconceptions can influence learning, we turn to the proposed neural education model of statistics cognition (Chapter 2) to frame how components of statistics cognition (literacy, reasoning, and thinking) may be differentially influences by misconceptions and biases.

For instance, consider the spacing effect, a cognitive bias that suggests that students can retain more information if they space their studying out as opposed to cramming information into a single study session the night before an exam (Crowder 1976; Greene, 1989; Ebbinghaus, 1964). Based on the NE model (Figure 2 of Chapter 2), the spacing effect seems most likely to negatively influence statistics literacy. That is, issues during encoding and retrieval may inhibit learning of statistics content. Although the spacing effect may have more to do with retrieval than encoding (Pyc & Rawson, 2009), knowledge of this effect is relevant when presenting materials relevant to that cognition. While the spacing effect can indirectly influence other statistics cognitions - after all, the NE model posits a specific relationship among all three features of statistics cognition - the processes behind statistics reasoning and thinking would need to be negatively influenced for this error to have a direct relationship (Table 2). While the answer to these sorts of questions are - at present - unknown, this is simply one example of how statistics educators can use knowledge of errors and content to troubleshoot statistics education.

This does not necessarily mean that educators must constantly screen students for potential biases, but there is an expectation that if a student is not performing up to their standards that the educator help the student ascertain the most likely reason and provide appropriate assistance. In fact, much of what an educator teaching an online course would do may occur before the first day of class, depending upon the pedagogical structure of their course (Beetham & Sharpe, 2013; Swan, 2001). We maintain that if educators take care to align their cognitively oriented curriculum with the appropriate frequency of cognitive objectives (Chapter 4), many of these issues can be resolved in advance.

Although biases can independently impact student outcomes, the NE model of statistics posits that educational outcomes share a hierarchical progression (Chapter 2). As such, cognitive biases could also pose simultaneous threats to different cognitive outcomes. For instance, the hindsight bias, that involves thinking that you knew something was predictable, after it occurred, can involve both memory (Mazzoni & Vannucci, 2007) and but can also falter reasoning skills, such as being able to adopt another's perspective (Birch & Bloom, 2007; Sommerville, Bernstein, & Meltzoff, 2013).

The biases briefly discussed throughout this book are not meant to be an exhaustive list, but rather a starting point of how models can be used to make testable predictions about cognitive biases and educational outcomes. By tracing biases to their cognitive roots and influenced neurological systems, scholars and educators may be able to understand how to curb their detrimental effects, and focus on what really matters in statistics education: improving student cognition.

Table 2. Examples of cognitive biases that could create misconceptions

Name	Subject	Type of Bias
Hindsight Bias	General	Reasoning and Literacy
Spacing Effect	General	Literacy
Availability heuristic	General	Reasoning

3. INFLUENCING STATISTICS COGNITIONS WITH EDUCATION

If cognitive biases are real (see Gigerenzer, 1996; Kahneman & Tversky, 1996), and can lead us astray, then we must acknowledge that to some degree we are not wired to be completely accurate when handling statistics information. This is problematic, as it suggests that people may have innate programming that they must overcome in order to become better statisticians. However, statistics education is oriented on cognitive development, meaning that cognitions are subject to change. Luckily, even if people are not intuitive statisticians (Kahneman & Tversky, 1982), there is some promising work that has suggested that cognitions processes - such as making predictions - can be improved (Mellers, Stone, Atanasov et al., 2015; Tetlock & Gardner, 2016).

To put this point in perspective, consider scope insensitivity, a classic cognitive bias that suggests people neglect to incorporate the multiplicative relationship of a valuation increasing in size (Desvouges, Naughton, & Parsons, 1992; Kahneman, 2000; Kahneman & Knetsch, 1992). In a typical paradigm, participants might be told that an effort is underway to clean up oil polluted lakes, with the exact number of lakes that will be cleaned to save a specified amount of birds varying based on random condition (Kahneman, Ritov, & Schkade, 1999). Upon being told how many birds would be saved (either 2,000; 20,000 or 200,000), participants were asked how much money they would donate to the cause. Regardless of condition, participants were willing to pay upwards of $88.

While scope insensitivity has been shown to be a pervasive cognitive bias, recent work has indicated that training may help people buffer against its negative influence (Tetlock & Gardner, 2016). For instance, in a series of studies conducted on forecasting tournaments sponsored by the US intelligence community (see Good Judgment Project, 2016), thousands of participants were asked rigorously vetted topical questions and asked to rate their likelihoods over time. While most people did indeed demonstrate a degree of scope neglect (Mellers, Stone, Murray, et al., 2015), a subset of the sample seemed to reduce the biases negative impact. When asked about the likelihood of a regime change in the Middle East, participants predicted that there was a 40% (after three months) or 41% (after six months) chance of a regime fall. In other words, most people ignore the time frame, and answered the easier question of whether or not they thought a regime change would occur. However, those who top performers, referred to as *super forecasters*

(Tetlock & Gardner, 2016), avoided some scope insensitivity by predicting 15% fall over three months, and a 24% fall over six months. Unlike typical forecasters, the super forecasters based their estimates on the time frame in the question.

These findings are not just important because they suggest that some individuals may be better at forecasting than others, but because they indicate that training, such as that provided to groups within the tournament, could improve the accuracy of forecasts. While super forecasters do appear have above average intelligence, Tetlock maintains that what makes a person a super forecaster is very much how they think about problems, and the amount of effort used to find relevant and accurate information (Tetlock & Gardner, 2016). In terms of statistics education, these results are meaningful in at least two ways. They suggest that some students may have more enhanced cognitions than others, needing different curriculum (Chapter 4), but also that even the top performers - such as super forecasters - can improve with training!

3.1. Developing Statistics Cognition

While you may be hard pressed to find an educator who does not value developing cognitions, the field of statistics education is more concerned with domain specific cognitions. As such, the NE model proposes connections between generalized cognitions, specified statistics cognitions, and educational outcomes. In one sense, these domain specific cognitions could be thought of as extensions of their discipline pure cognitive process. That is, statistics reasoning is derived from a generalized reasoning ability. However, even if a student has advanced generalized cognitions, it does not necessarily suggest that they are any more able to be successful in developing their statistics cognitions, relative to peers without generalized cognitive advantages. As previously discussed, part of learning statistical operations has to do with one's ability to appropriately allocate resources toward controlled processes.

Given that cognitions are dynamic - as opposed to static - processes, the key to improving cognitions is by structuring education to be developmental, measuring change over time. Indeed, developing statistics cognitions has been of central interest to researchers in statistics education for some time (Ben-Zvi & Garfield, 2004; Chance, Garfield, & delMas, 1999; Wild & Pfannkuch, 1999). Whether the target aim was to improve literacy (Gal, 2004), reasoning (Garfield, 2002), or thinking (Pfannkuch & Wild, 2004), scholars have identified systematic methods that have proven successful in enhancing cognitive development. While much of the research aimed at developing student cognition attempts to uniquely identify and improve literacy, reasoning, and thinking, the NE model demands a holistic approach to measuring these cognitions in relation to one another, and other psychological variables.

Because cognitive learning objectives may not lend themselves to traditional modes of assessment as readily as behavioral objectives (see Chapter 4), scholars have proposed that active learning techniques be the primary vehicle used to spur cognitive growth (Mayer, 2004; Prince, 2004). However, there has been mixed success in understanding the role active learning plays for student outcomes. For example, Pfaff and Weinberg (2009) employed a case study approach to assess the efficacy of using activities to increase student understanding. They found that despite great effort, student learning did not improve as a result of these activities relative to non-active approaches.

Although Pfaff and Weinberg (2009) did not find evidence supporting the use of active learning practices, other studies have suggested that that the success of active learning depends on the type of student (Weltman & Whiteside 2010). In their experiment, Weltmann and Whiteside compared over 300 statistics courses offered within business departments at a higher educational institution. Researchers randomly assigned methods and content, to ensure that all students were received active, traditional (lecturing), and hybrid pedagogical approaches. When broken into groups based on their GPA, Weltman and Whiteside noted that students with low GPAs tended to perform better as a function of active learning. Conversely, students with high GPAs tended to drop to the mean performance when exposed to active learning. Although there are a variety of ways these results could be interpreted, students with low GPAs could lack confidence, motivation, or the base level knowledge needed to succeed.

By comparing the methodologies from these studies, the NE model of statistics cognition proposes several features of active learning practices that educators must be mindful of. First, in order to determine whether or not active learning works, it is necessary to specify to what end active learning is supposed to work. That is, why is the active learning being used, what benefit does it afford over traditional lecturing, and how is it tailored to enhance cognition? Although it is difficult to measure and assess cognitive development, educators must consistently put their methods to the test and collect evidence that sheds light on whether or not their method is working, and how it could be improved for the future. Second, if active learning is about cognitive growth, it stands to reason that it should only be employed when there is a chance for such growth for all students. To further illustrate these concepts, consider each study revisited.

First, Pfaff and Weinberg (2009) explicitly state that they are attempting to assess whether active learning increases understanding. However, to assess an increase in understanding, the authors used several exams consisting of multiple choice and short answer questions. This may be problematic, for as discussed (see Chapter 4), it remains unclear whether or not the items on these exams all assessed student understanding, and to what degree these exams were measuring cognitive growth. Arguably, some of these items assess ability to recall information, such as multiple choice items asking for definitions ("A 95% confidence interval indicates that..."), with others geared more toward conceptual reasoning ("Which graph represents a distribution of sample means for 500 samples of size 4?"). Importantly, with the NE model, cognitive abilities are linked to educational outcomes, that are different based on the specific cognition that the educational assessment is designed to assess (see Figure 3 of Chapter 4). Indeed, Pfaff and Weinberg note that some items seem to suggest improved learning, but others do not. If this data made use of the NE model of statistics cognition, it is possible that Pfaff and Weinberg could find that specific educational outcomes, such as questions relating to understanding, demonstrated wider improvements when active learning was used.

Further, any effect within the context of Pfaff and Weinberg's (2009) case study may be hidden by differences in student ability. As previously mentioned, low GPA students seem to benefit more from active learning than higher GPA students. This could suggest that higher GPA students have an innate difference that requires a different set of curriculum (Chapter 4), but could also suggest that active learning is used to a specific cognitive end. That is, to master a specific skill or process. Instead of using active learning to teach definitions and help students recall information, students with higher GPAs may benefit from targeted active learning that is explicitly designed to use meta-cognitive skills.

For example, when students are unsure of a concept, an activity that provides hands on experiences or demonstrates specific connections among concepts is likely to be both memorable and of educational value. However, if students are already aware of many of the concepts, the activity does not present any sort of challenge, that could lead to students disengaging (Dai, Moon, &

Feldhusen, 1998; Feldhusen & Kroll, 1991; Gallagher, Harradine, Coleman, 1997). However, familiarity of the concept is not the only characteristic of interest, as the NE model of statistics cognition also supports a hierarchical structure of information. That is, for an activity to be effective, it must be designed in a way that allows student - regardless of ability - a chance to learn in an effective and efficient manner.

By way of example, having a debate about key terms and definitions can be useful, and may even be enjoyable, but what matters most is whether or not it is an efficient and effective means to develop cognition, that must be verified with evidence. Although there is certainly need to assess and understand difference among student affect, styles of learning may be more preferential than they are predictors of the way we learn (Chapter 6). As previously suggested, what determines whether an activity is meaningful or efficient has a lot to do with the objective the activity is aimed at facilitating, that can be neatly classified within the context of the NE model of statistics cognition.

Further, because the NE model of education predicts specific relationships among types of statistics cognition, it suggests that active learning practices result in optimal student outcomes when they adhere to the theorized structural progression of knowledge. In other words, an activity designed to help students understand (reason) is of little use if students are unable to identify or recall key terms (literacy). That is not to say that active learning cannot be used for both reasoning and literacy purposes, but rather the structure of the active learning should vary depending upon comfort and familiarity students have with the topic. Largely, this about placing students in their zone of proximal development (Vygotsky, 1987).

Scholars are often looking for ways to make their classes more enjoyable, but an educators first responsibility is to ask whether the activity is the most efficient and effective way of learning the material. In cases that classes are built from a cognitive curriculum, educators must continuously assess both content and skills. As will be discussed (see Chapter 4), this is largely achieved by considering what the learning goal for the students is, and what types of tasks are required to assess that goal. Above all else, these activities need to be framed on cognitive, as opposed to behavioral, outcomes (Mayer, 2004). This means that each activity must explicitly state the way the activity will demonstrate growth of cognitive processes.

3.2. Developing Cognitive Outcomes in Online Environments

Active learning practices are simply alternative methods to educating students. These methods are deemed alternative, as they go beyond requiring students to listen to a lecture. While there could be a premium on using active learning strategies over formal lecture practices, even students readily acknowledge that a large part of learning mathematics is rote memorization of formula (Crawford, Gordon, Nicholas, & Prosser, 1994). That is not to say that active learning techniques, such as asking students to generate examples, cannot be integrated into curriculum, but that at some level students will need repeated exposure to obtain proficiency.

Although the research previously outlined focuses on active learning in face-to-face environments, many of the constraints discussed apply to online environments (see Mehlenbacher, Miller, Covington, and Larsen, 2000; Salmon, 2013). However, active learning is generally associated with assessing behavioral objectives, and as previously discussed, the objectives in a statistics classroom be framed on developing cognitions. Generally, to develop cognitive outcomes in an online environments, scholars have focused on developing communities of inquiry (Garrison, Anderson, & Archer, 2000). To achieve a community of inquiry, educators design courses that generate feelings of unease from their educational experiences. This is argued to motivate students as they seek out information that can lessen their dissonance. As the student assesses ideas and makes connections, they can apply what they have learned to reach an appropriate conclusion.

Using a focus group methodology, researchers aimed to determine how developing this cognitive presence could facilitate higher order learning (Kanuka & Garrison, 2004). In their study, Kanuka and Garrison identified external (discourse, discourse, management) and internal (reflection, self-monitoring, information processing) features required to create cognitive presence. While creating cognitive presence has been shown to benefit learning (Akyol & Garrison, 2011; Garrison et al., 2001; Vaughan & Garrison, 2005), activities must be well structure, provide clearly defined roles and responsibilities, and provoke students to confront one another to be of value (Kanuka, Rourke, & Laflamme, 2007). Thus, developing cognitive outcomes in statistics education may be relatively similar, regardless of modality.

4. ONLINE LEARNING IMPLICATIONS: ENHANCING STUDENT COGNITION

Although studies may not demonstrate differences in terms of student performance, there is still a need to consider how cognitions can uniquely interact with online environments. From a general cognitive perspective, there may be several theoretical ways in that college student cognition could interact with their online learning environments. As previous discussed in Chapter 2, statistics cognitions can be related to general cognitions, and student outcomes, and so one could anticipate that by improving cognition they can improve statistics cognition, but this may not always be true, as one cannot become a statistical expert without *deliberate practice* (Ericsson, Krampe, & Tesch-Römer, 1993). Yet because these general cognitions can have an impact of specialized cognitive processing, it is necessary to consider how pre-existing cognitive structures could impact learning in digital environments.

For example, if students are cognitive misers they are geared toward saving as much cognitive resource as possible. As a result, one of the first things a student enrolled in an online course may examine what types of assignments they have, and then decide the type of approach to their work. For example, consider a student in an online class that has four open book exams, each with a one week time limit. While this exam, if constructed properly, will not allow students any more affordances to cheat (Rowe, 2004), it may signal to the student that they do not necessarily need to work on reading all information as carefully, and may prompt them to treat the internet as an online repository for information that they will seek as needed. Conversely, establishing a timed selected response exam in an online environment is more similar to the type of assessments students in face-to-face courses may be exposed to, suggesting that in this instance, students may employ a similar study strategy as to the ones they used in online courses.

Further, one can consider the role of planning in an online course, and how it differs from that of a face-to-face classroom. Again, depending on the number, and types, of assessments, students may modify the amount of time they spend in the online course. Yet, regardless of time commitment, online courses pose some additional considerations with planning. For example, at any moment, students could lose access to their course materials. Whether the outage is due to a home, or University technical issue, such a problem is often unexpected and could mean the difference in terms of whether or not an assignment is on time, or late. Further, because online courses often involve web 2.0 tools (see Conole & Alevizou, 2010), online courses may often have specific learning objectives that have less to do with discipline specific course content, and more to do with demonstrating an ability to make use of a specific computer program. For instance, many statistics courses may encourage students to use Excel™ (Warner & Meehan, 2001), so that students can demonstrate a command of data through its manipulation. However, even implicitly, classes may often assume students have technical knowhow and the ability to locate and make use of credible sources from the internet. If web 2.0 tools are to be used as a critical components of the course (i.e., all assignments are done in Excel™), specific learning objectives need to be structured to ensure that students are literate and prepared to reason and think about statistics in the context of this program. Students who experience computer anxiety may recognize that learning this program could take a considerable amount of time, and plan their work differently. In essence, while the processes students learn can be translated into an online environment, online courses may be at a content-specific disadvantage, as using web 2.0 tools may inherently become part of course learning objectives.

Failure to incorporate tool-specific learning objectives, such as "students will demonstrate how to use Excel™ to manipulate data", could create undue frustration, and prompt students without a software background to feel concerned. Indeed, many of the differences in terms of online education seem affective in nature, regardless if there is a direct impact on student performance (see Bernard et al., 2004). For example, students in online courses often report that educators do not know them personally (El Mansour & Mupinga, 2007), that could indirectly influence their performance if students interpret a lack of interaction as a low degree of instructor availability (Moore & Kearsley, 1996). If students do not feel that they receive appropriate feedback (Finaly-Neumann, 1994) they may have less satisfaction with online learning, that can be a strong indicator of future performance in online coursework (DeBourgh, 1999; Hiltz, 1993). Further, if students perceive that the instructor is not communicating with them regularly (Mood, 1995), they may become frustrated (Hara & Kling, 2000), and disengage from learning. Again, while many of the issues reviewed are not necessarily unique to online they are still relevant to crafting an ideal learning environment. Indeed, many of the obstacles common to face-to-face learning may just manifest in different ways in online environments.

For instance, face-to-face courses meet at consistent times, with professors often mandated to hold a specified amount of office hours at a certain time on a certain day. In an online environment, instructors may make use of the flexibility, constructing time specified policies, such as stating students will receive a response within 48 hours. This again becomes a problem for a student who does not plan accordingly, and runs into a problem the night before an assignment is due. While the same problem could occur in a face-to-face class, the student may not ever see the educator in an online course, and the boundaries in terms of what is fair and when responses can be expected may be less clear.

However, scholars in online environments are aware of these important nuances, and have long since discussed ways that various cognitions can be developed in online environments (Govindasamy, 2001; Steinkuehler, 2008). Indeed, scholars in online educational research are well aware of the psychological challenges in constructing online courses, often suggesting ways educators can develop social (Swan & Shih, 2005), cognitive (Garrison & Cleveland-Innes, 2005), and teaching (Shea, Li, Pickett, 2006) presence in online classes. Yet, regardless of the exact valence, it seems clear that our general cognitions can be influenced by the nature of the learning environment one finds themselves in, regardless as to whether the mode of instructional delivery is digital or in person (Garrison, 2003; Herrington, Oliver, & Reeves, 2003; Hung & Chen, 2001; Stahl, 2006).

Central to building a statistics course in an online environment is considering how the materials will be used by students. For example, if online courses act as an information repository for students, they may be more selective with the extent and depth of their reading of assigned material. Indeed, the use of technology has been shown to have a direct impact on memory systems (James & Engelhardt, 2012; Sana, Weston, & Cepeda, 2013), that are crucial to developing statistics literacy. For instance, in one study, researchers examined classroom performance and found that students who took handwritten notes tended to recall the material better than peers who took notes on a computer (Mueller & Oppenheimer, 2014).

Importantly, those on computers were not simply distracted, in fact, students who used laptops took copious amounts of notes, relative to peers taking notes on paper. Mueller and Oppenheimer interpret their results to suggest that that students who used computers took notes in a mindless fashion, copying each word verbatim. The act of writing notes may be advantageous to students, yet in an online classroom, students may find little need to even take notes. Even those who take notes may be more likely to use a computer to complete this task given that they may be more likely to be familiar with technology (Levin & Arafeh, 2002; Vie, 2008), or simply based on the fact that they are required to use a computer to access their course material, and select this mode of note taking as a convenience.

That is not to say that students who use using technology negatively influences memory or learning. Indeed, these results might best be explained by the effort that goes into the act of note taking (Piolat, Olive, & Kellog, 2005) as a function of modality. For example, taking notes can act as an external place to store information so that we do not have to encode or retrieve it at a moment's notice (Rickards & Friedman, 1978). However, note taking can also be structured in a way that provides the learner with targeted cues, and helps them visualize the organizational relationship among terms (Kiewra, 1985). In the previously described study, the authors asserted that while technology allowed students to take more notes, quantity was trumped by quality (James & Engelhardt, 2012). This serves as an important reminder, as technology may often make learning easier for students; however, we must be certain that it does not do the learning for the student.

This is but one example of how cognition can be influenced by technology. Much of what needs to be considered depends greatly on the specific learning objectives selected for a course. For example, a course that focuses most of its learning objectives on memorizing content may signal students that they should take notes as a primary means to succeed. Conversely, a course that relies heavily on critical thinking may require some note taking, but self-elaboration and other higher order learning strategies are of greater value for these sorts of tasks (Dunlosky et al., 2013). What may be most critical in designing a course based on cognitive objectives is determining the best ways for students to demonstrate that their literacy, reasoning, and thinking have improved. In an online environment, this means carefully selecting web 2.0 tools that ease the burden of rote learning, while forcing students to experience dissonance and creatively apply their knew knowledge to viable solutions.

5. SUMMARY: ENHANCING COGNITIVE OUTCOMES

This Chapter focused on two central themes. The first had to do with defining and laying out a framework for understanding the operational processes of statistics cognitions. What sets developing these cognitive processes apart from *typical* learning experiences, is that it pays close attention to the distinct sources of biases and misconceptions when troubleshooting student errors. We have argued that depending on the source, errors can be attributed to different processes, such as motivation, or self-regulation, that in turn prompt the educator to attempt different solutions.

The second theme focused on developing statistics cognition. When we talk about development, we are referring to the continuous measurement of a dynamic process to examine changes over time. Although developmental research is temporally aligned, not all research that includes time series can be considered developmental. What really sets developmental research apart is its dedication to understand does something happen, as opposed to simply answer if something can happen (McCall, 1977).

Given that student outcomes need not differ by modality, one may ask whether there is a need to make a distinction between learning in face-to-face or online environment. While most students can perform just as well in either environment, performance is only one of many possible student outcomes, and may not be the most appropriate indicator of learning in higher education environments. The limited studies ($n = 47$) we identified as assessing statistics in online environments may even be subject to a lack of generalization if students in face-to-face and online classes are systematically different due to a self-selection bias (see Dutton, Dutton, & Perry, 2002). While many researchers attempt to circumvent the fact that any single classroom may not necessarily have selected students randomly by randomly selecting classrooms and schools (Schochet, 2007), it remains unclear to what degree consistent individual differences between face-to-face and online student populations emerge.

For instance, face-to-face course completion rates tend to be 10 - 20% higher than those of online courses (Carr, 2000). A finding Carr interpreted as differences attributed to student expectations, noting that many online students work full time careers and have other priorities, such as family whom they care for. This divide could even be exacerbated by the finding that online institutions tend to have cheaper tuition (Deming, Goldin, Katz, & Yutchman, 2015), suggesting possible differences in socioeconomic status among modes of learning. However, what is clear is that despite possible self-selection biases, students who enroll in online courses perform just as well, if not better, than peers enrolled in face-to-face courses (Allen et al., 2004; Cavanaugh, Gillan, Komrey, Hess, & Blomeyer, 2004).

While we acknowledge that we may see differences in performance among students based on the modality of instruction, we concur with prior research that finds that the diversity of types of instruction in both online and face-to-face formats makes it exceptionally difficult to measure whether differences in learning emerge. Based on the NE model, it can also be asserted that performance is strictly defined as cognitive enhancement, in that a statistics course means demonstrated improvement of literacy, reasoning, and thinking. Unfortunately, without knowing how well performance in any given course was linked to these cognitive outcomes, it is difficult to say whether or not cognitive differences emerge in online learning environment.

If humans are indeed hardwired to interpret and make use of statistical information, then it becomes clear that teaching statistics is teaching for cognition. All students are capable, to some degree, of retaining information and thinking critically about ideas. Yet more work needs to focus on ways to understand how automated thinking may lead students to a consistent pattern of errors in online environments. This Chapter demonstrated the importance of why linking educational errors to cognitive processes is necessary, and provided some evidence as to how technology can influence cognitive systems. It is time to test specific theories, and identify structural ways online courses can be adapted to intuitively facilitate literacy, reasoning, and thinking.

REFERENCES

Akyol, Z., & Garrison, D. R. (2011). Understanding cognitive presence in an online and blended community of inquiry: Assessing outcomes and processes for deep approaches to learning. *British Journal of Educational Technology*, *42*(2), 233–250. doi:10.1111/j.1467-8535.2009.01029.x

Allen, M., Mabry, E., Mattrey, M., Bourhis, J., Titsworth, S., & Burrell, N. (2004). Evaluating the effectiveness of distance learning: A comparison using meta-analysis. *Journal of Communication*, *54*(3), 402–420. doi:10.1111/j.1460-2466.2004.tb02636.x

Allport, A. (1993). Attention and control: Have we been asking the wrong questions? A critical review of twenty-five years. In D.E. Meyer & S. Kornblum (Eds.), *Attention and performance XIV: Synergies in experimental psychology, artificial intelligence, and cognitive neuroscience* (pp. 183-218). Cambridge, MA: A Bradford Book.

Ayton, P., & Fischer, I. (2004). The hot hand fallacy and the gamblers fallacy: Two faces of subjective randomness?. *Memory & Cognition*, *32*(8), 1369–1378. doi:10.3758/BF03206327 PMID:15900930

Baddeley, A. (2003). Working memory: Looking back and looking forward. *Nature Reviews. Neuroscience*, *4*(10), 829–839. doi:10.1038/nrn1201 PMID:14523382

Batanero, C., Godino, J. D., Vallecillos, A., Green, D. R., & Holmes, P. (1994). Errors and difficulties in understanding elementary statistical concepts. *International Journal of Mathematical Education in Science and Technology*, *25*(4), 527–547. doi:10.1080/0020739940250406

Baumeister, R. F., & Heatherton, T. F. (1996). Self-regulation failure: An overview. *Psychological Inquiry*, *7*(1), 1–15. doi:10.1207/s15327965pli0701_1

Baumeister, R. F., Vohs, K. D., & Tice, D. M. (2007). The strength model of self-control. *Current Directions in Psychological Science*, *16*(6), 351–355. doi:10.1111/j.1467-8721.2007.00534.x

Beetham, H., & Sharpe, R. (2013). *Rethinking pedagogy for a digital age: Designing for 21st century learning*. New York, NY: Routledge.

Ben-Zvi, D., & Garfield, J. B. (2004). Introduction to statistical literacy, reasoning, and thinking. In D. Ben-Zvi & J. B. Garfield (Eds.), *The challenge of developing statistical literacy, reasoning and thinking* (pp. 1–17). Dordrecht, Netherlands: Kluwer Academic Publishers. doi:10.1007/1-4020-2278-6

Bernard, R. M., Abrami, P. C., Lou, A. Y., Borokhovski, A. W., Wade, L., Wozney, L., & Huang, B. et al. (2004). How does distance education compare with classroom instruction? A meta-analysis of the empirical literature. *Review of Educational Research*, *74*(3), 379–439. doi:10.3102/00346543074003379

Birch, S. A., & Bloom, P. (2007). The curse of knowledge in reasoning about false beliefs. *Psychological Science*, *18*(5), 382–386. doi:10.1111/j.1467-9280.2007.01909.x PMID:17576275

Bolliger, D. U., Supanakorn, S., & Boggs, C. (2010). Impact of podcasting on student motivation in the online learning environment. *Computers & Education*, *55*(2), 714–722. doi:10.1016/j.compedu.2010.03.004

Borst, J. P., Taatgen, N. A., & van Rijn, H. (2010). The problem state: A cognitive bottleneck in multitasking. *Journal of Experimental Psychology. Learning, Memory, and Cognition*, *36*(2), 363–382. doi:10.1037/a0018106 PMID:20192536

Bowman, L. L., Levine, L. E., Waite, B. M., & Gendron, M. (2010). Can students really multitask? An experimental study of instant messaging while reading. *Computers & Education*, *54*(4), 927–931. doi:10.1016/j.compedu.2009.09.024

Brown, P. C., Roediger, H. L., & McDaniel, M. A. (2014). *Make it stick*. Cambridge, MA: Harvard University Press. doi:10.4159/9780674419377

Carr, S. (2000). As distance education comes of age, the challenge is keeping the students. *The Chronicle of Higher Education*, *46*, 39–41.

Carver, C. S., & Scheier, M. F. (1988). A control-process perspective on anxiety. *Anxiety Research*, *1*(1), 17–22. doi:10.1080/10615808808248217

Cavanaugh, C., Gillan, K. J., Kromrey, J., Hess, M., & Blomeyer, R. (2004). *The effects of distance education on K-12 student outcomes: A meta-analysis*. Naperville, IL: Learning Point Associates/North Central Regional Educational Laboratory (NCREL).

Chance, B., Garfield, J., & delMas, B. (1999). *A model of classroom research in action: Developing simulation activities to improve students' statistical reasoning*. 52nd Session of the International Statistical Institute, Helsinki, Finland.

Conole, G., & Alevizou, P. (2010). *A literature review of the use of Web 2.0 tools in Higher Education*. York, UK: Higher Education Academy.

Crawford, K., Gordon, S., Nicholas, J., & Prosser, M. (1994). Conceptions of mathematics and how it is learned: The perspectives of students entering university. *Learning and Instruction*, *4*(4), 331–345. doi:10.1016/0959-4752(94)90005-1

Crowder, R. G. (1976). *Principles of learning and memory*. Hillsdale, NJ: Lawrence Erlbaum Associates.

Cumming, G., Williams, J., & Fidler, F. (2004). Replication and researchers understanding of confidence intervals and standard error bars. *Understanding Statistics*, *3*(4), 299–311. doi:10.1207/s15328031us0304_5

Dai, D. Y., Moon, S. M., & Feldhusen, J. F. (1998). Achievement motivation and gifted students: A social cognitive perspective. *Educational Psychologist*, *33*(2-3), 45–63. doi:10.1080/00461520.1998.9653290

Debourgh, G. A. (1999). *Technology is the tool, teaching is the task: Student satisfaction in distance learning*. Chesapeake, VA: Association for the Advancement of Computing in Education (AACE).

Deming, D. J., Goldin, C., Katz, L. F., & Yuchtman, N. (2015). Can online learning bend the higher education cost curve?. *The American Economic Review*, *105*(5), 496–501. doi:10.1257/aer.p20151024

Desvouges, W. H., Naughton, M. C., & Parsons, G. R. (1992). Benefits transfer: Conceptual problems in estimating water quality benefits using existing studies. *Water Resources Research*, *28*(3), 675–683. doi:10.1029/91WR02592

Dewey, D. R., & Keynes, J. N. (1891). Review: Keynes on statistics: The scope and method of political economy by John Neville Keynes. *Publications of the American Statistical Association*, *2*(14), 308–310. doi:10.2307/2276579

Dijksterhuis, A., Bos, M. W., Nordgren, L. F., & Van Baaren, R. B. (2006). On making the right choice: The deliberation-without-attention effect. *Science*, *311*(5763), 1005–1007. doi:10.1126/science.1121629 PMID:16484496

Dijksterhuis, A., & Meurs, T. (2006). Where creativity resides: The generative power of unconscious thought. *Consciousness and Cognition*, *15*(1), 135–146. doi:10.1016/j.concog.2005.04.007 PMID:16019229

Dijksterhuis, A., & Nordgren, L. F. (2006). A theory of unconscious thought. *Perspectives on Psychological Science*, *1*(2), 95–109. doi:10.1111/j.1745-6916.2006.00007.x PMID:26151465

Dougherty, M. R., Gettys, C. F., & Ogden, E. E. (1999). MINERVA-DM: A memory processes model for judgments of likelihood. *Psychological Review*, *106*(1), 180–209. doi:10.1037/0033-295X.106.1.180

Dunlosky, J., Rawson, K. A., Marsh, E. J., Nathan, M. J., & Willingham, D. T. (2013). Improving students learning with effective learning techniques: Promising directions from cognitive and educational psychology. *Psychological Science in the Public Interest*, *14*(1), 4–58. doi:10.1177/1529100612453266 PMID:26173288

Dutton, J., Dutton, M., & Perry, J. (2002). How do online students differ from lecture students. *Journal of Asynchronous Learning Networks*, *6*, 1–20.

Ebbinghaus, H. (1964). *Memory: A contribution to experimental psychology* (H.A. Ruger, C.E. Bussenius, & E.R. Hilgard, Trans.). New York, NY: Dover Publications, Inc.

El Mansour, B., & Mupinga, D. M. (2007). Students' Positive and Negative Experiences in Hybrid and Online Classes. *College Student Journal, 41*, 242–248.

Ericsson, K. A., Krampe, R. T., & Tesch-Römer, C. (1993). The role of deliberate practice in the acquisition of expert performance. *Psychological Review, 100*(3), 363–406. doi:10.1037/0033-295X.100.3.363

Feldhusen, J. F., & Kroll, M. D. (1991). Boredom or challenge for the academically talented in school. *Gifted Education International, 7*(2), 80–81. doi:10.1177/026142949100700207

Field, A. (2013). *Discovering statistics using IBM SPSS statistics* (3rd ed.). Thousand Oaks, CA: Sage Publications.

Finaly-Neumann, E. (1994). Course work characteristics and students' satisfaction with instruction. *Journal of Instructional Psychology, 21*, 14–24.

Fiske, S. T., & Taylor, S. E. (2013). *Social cognition: From brains to culture*. Thousand Oaks, CA: Sage. doi:10.4135/9781446286395

Gailliot, M. T., & Baumeister, R. F. (2007). The physiology of willpower: Linking blood glucose to self-control. *Personality and Social Psychology Review, 11*(4), 303–327. doi:10.1177/1088868307303030 PMID:18453466

Gal, I. (2004). Statistical literacy. In D. Ben-Zvi & J. B. Garfield (Eds.), *The challenge of developing statistical literacy, reasoning and thinking* (pp. 47–78). Dordrecht, Netherlands: Kluwer Academic Publishers. doi:10.1007/1-4020-2278-6_3

Gallagher, J., Harradine, C. C., & Coleman, M. R. (1997). Challenge or boredom? Gifted students views on their schooling. *Roeper Review, 19*(3), 132–136. doi:10.1080/02783199709553808

Gardenier, J., & Resnik, D. (2002). The misuse of statistics: Concepts, tools, and a research agenda. *Accountability in Research: Policies and Quality Assurance, 9*(2), 65–74. doi:10.1080/08989620212968 PMID:12625352

Garfield, J.B. (1995). How students learn statistics. *International Statistical Review/Revue Internationale de Statistique, 63*, 25-34. doi: 10.2307/1403775

Garfield, J. B. (2002). The challenge of developing statistical reasoning. *Journal of Statistics Education, 10*, 58–69.

Garfield, J. B. (2003). Assessing statistical reasoning. *Statistics Education Research Journal, 2*, 22–38.

Garfield, J. B., & Ahlgren, A. (1988). Difficulties in learning basic concepts in probability and statistics: Implications for research. *Journal for Research in Mathematics Education, 19*(1), 44–63. doi:10.2307/749110

Garfield, J. B., & Ben-Zvi, D. (2007). How students learn statistics revisited: A current review of research on teaching and learning statistics. *International Statistical Review, 75*(3), 372–396. doi:10.1111/j.1751-5823.2007.00029.x

Garrison, D. R. (2003). Cognitive presence for effective asynchronous online learning: The role of reflective inquiry, self-direction and metacognition. In J. Bourne & J. C. Moore (Eds.), *Elements of quality online education: Practice and direction, 4* (pp. 47–58). Needham, MA: Sloan-C.

Garrison, D. R., Anderson, T., & Archer, W. (2001). Critical thinking, cognitive presence, and computer conferencing in distance education. *American Journal of Distance Education, 15*(1), 7–23. doi:10.1080/08923640109527071

Garrison, D. R., & Cleveland-Innes, M. (2005). Facilitating cognitive presence in online learning: Interaction is not enough. *American Journal of Distance Education, 19*(3), 133–148. doi:10.1207/s15389286ajde1903_2

Gigerenzer, G. (1996). On narrow norms and vague heuristics: A reply to Kahneman and Tversky. *Psychological Review, 103*(3), 592–596. doi:10.1037/0033-295X.103.3.592

Good Judgment Project. (2016). Retrieved from http://www.goodjudgment.com/

Govindasamy, T. (2001). Successful implementation of e-learning: Pedagogical considerations. *The Internet and Higher Education, 4*(3-4), 287–299. doi:10.1016/S1096-7516(01)00071-9

Greene, R. L. (1989). Spacing effects in memory: Evidence for a two-process account. *Journal of Experimental Psychology. Learning, Memory, and Cognition*, *15*(3), 371–377. doi:10.1037/0278-7393.15.3.371

Hagger, M. S., Wood, C., Stiff, C., & Chatzisarantis, N. L. (2010). Ego depletion and the strength model of self-control: A meta-analysis. *Psychological Bulletin*, *136*(4), 495–525. doi:10.1037/a0019486 PMID:20565167

Hara, N., & Kling, R. (2000). Students distress with a web-based education course: An ethnographic study of participants experiences. *Information Communication and Society*, *3*, 557–579. doi:10.1080/13691180010002297

Harris, K. R., & Graham, S. (1985). Improving learning disabled students composition skills: Self-control strategy training. *Learning Disability Quarterly*, *8*(1), 27–36. doi:10.2307/1510905

Haselton, M. G., Nettle, D., & Andrews, P. W. (2005). The evolution of cognitive bias. In D. M. Buss (Ed.), *The handbook of evolutionary psychology* (pp. 724–746). Hoboken, NJ: John Wiley & Sons Inc.

Hembrooke, H., & Gay, G. (2003). The laptop and the lecture: The effects of multitasking in learning environments. *Journal of Computing in Higher Education*, *15*(1), 46–64. doi:10.1007/BF02940852

Herrington, J., Oliver, R., & Reeves, T. C. (2003). Patterns of engagement in authentic online learning environments. *Australasian Journal of Educational Technology*, *19*(1), 59–71. doi:10.14742/ajet.1701

Hiltz, S. R. (1993). Correlates of learning in a virtual classroom. *International Journal of Man-Machine Studies*, *39*(1), 71–98. doi:10.1006/imms.1993.1054

Hirsch, L. S., & O'Donnell, A. M. (2001). Representativeness in statistical reasoning: Identifying and assessing misconceptions. *Journal of Statistics Education*, *9*, 61–82. Retrieved from https://ww2.amstat.org/publications/jse/v9n2/hirsch.html

Hodges, C. B. (2004). Designing to motivate: Motivational techniques to incorporate in e- learning experiences. *Journal of Interactive Online Learning*, *2*, 1–7.

Horn, L., Peter, K., Rooney, K., & Malizio, A. G. (2002). *Profile of undergraduates in US postsecondary institutions: 1999-2000*. Washington, DC: US Department of Education, Office of Educational Research and Improvement, National Center for Education Statistics.

Hung, D. W., & Chen, D. T. (2001). Situated cognition, Vygotskian thought and learning from the communities of practice perspective: Implications for the design of web-based e- learning. *Educational Media International*, *38*(1), 3–12. doi:10.1080/09523980121818

James, K. H., & Engelhardt, L. (2012). The effects of handwriting experience on functional brain development in pre-literate children. *Trends in Neuroscience and Education*, *1*(1), 32–42. doi:10.1016/j.tine.2012.08.001 PMID:25541600

Kahneman, D. (2000). Evaluation by moments: Past and future. In D. Kahneman & A. Tversky (Eds.), *Choices, Values, and Frames* (pp. 693–708). New York, NY: Cambridge University Press and the Russell Sage Foundation.

Kahneman, D. (2011). *Thinking, fast and slow*. New York, NY: Macmillan.

Kahneman, D., & Klein, G. (2009). Conditions for intuitive expertise: A failure to disagree. *The American Psychologist*, *64*(6), 515–526. doi:10.1037/a0016755 PMID:19739881

Kahneman, D., & Knetsch, J. L. (1992). Valuing public goods: The purchase of moral satisfaction. *Journal of Environmental Economics and Management*, *22*(1), 57–70. doi:10.1016/0095-0696(92)90019-S

Kahneman, D., Ritov, I., & Schkade, D. (1999). Economic preferences or attitude expressions?: An analysis of dollar responses to public issues. *Journal of Risk and Uncertainty*, *19*(1/3), 203–235. doi:10.1023/A:1007835629236

Kahneman, D., Slovic, P., & Tversky, A. (1982). *Judgment under uncertainty: Heuristics and biases*. Cambridge, UK: Cambridge University Press. doi:10.1017/CBO9780511809477

Kahneman, D., & Tversky, A. (1973). On the psychology of prediction. *Psychological Review*, *80*(4), 237–251. doi:10.1037/h0034747

Kahneman, D., & Tversky, A. (1982). On the study of statistical intuitions. *Cognition, 11*(2), 123–141. doi:10.1016/0010-0277(82)90022-1 PMID:7198957

Kahneman, D., & Tversky, A. (1996). On the reality of cognitive illusions. *Psychological Review, 103*(3), 582–591. doi:10.1037/0033-295X.103.3.582 PMID:8759048

Kanuka, H., & Garrison, D. R. (2004). Cognitive presence in online learning. *Journal of Computing in Higher Education, 15*(2), 21–39. doi:10.1007/BF02940928

Kanuka, H., Rourke, L., & Laflamme, E. (2007). The influence of instructional methods on the quality of online discussion. *British Journal of Educational Technology, 38*(2), 260–271. doi:10.1111/j.1467-8535.2006.00620.x

Karpicke, J. D., Butler, A. C., & Roediger, H. L. III. (2009). Metacognitive strategies in student learning: Do students practise retrieval when they study on their own?. *17*(4), 471–479. doi:10.1080/09658210802647009 PMID:19358016

Kiewra, K. A. (1985). Investigating notetaking and review: A depth of processing alternative. *Educational Psychologist, 20*(1), 23–32. doi:10.1207/s15326985ep2001_4

Kunda, Z., & Nisbett, R. E. (1986). Prediction and the partial understanding of the law of large numbers. *Journal of Experimental Social Psychology, 22*(4), 339–354. doi:10.1016/0022-1031(86)90019-3

Kundert, K. (1990). Student-generated data in elementary statistics. *Mathematics Teacher, 83*, 322–325.

Lavasani, M. G., Weisani, M., & Shariati, F. (2014). The role of achievement goals, academic motivation in statistics anxiety: Testing a causal model. *Procedia: Social and Behavioral Sciences, 114*, 933–938. doi:10.1016/j.sbspro.2013.12.810

Lehrer, J. (2009). *How We Decide*. New York, NY: Houghton Mifflin Harcourt.

Levin, D., & Arafeh, S. (2002). *The digital disconnect: The widening gap between Internet-savvy students and their schools*. Washington, DC: Pew Internet & American Life Project.

Li, K. Y., & Shen, S. M. (1992). Students weaknesses in statistical projects. *Teaching Statistics*, *14*(1), 2–9. doi:10.1111/j.1467-9639.1992.tb00195.x

Lovett, M. C., & Greenhouse, J. B. (2000). Applying cognitive theory to statistics instruction. *The American Statistician*, *54*, 196–206. doi:10.1080/00031305.2000.10474545

Magel, R. C. (1996). Increasing student participation in large introductory statistics classes. *The American Statistician*, *50*, 51–56. doi:10.1080/00031305.1996.10473542

Mallows, C. (1998). The Zeroth Problem. *The American Statistician*, *52*, 1–9.

Mayer, R. E. (2004). Should there be a three-strikes rule against pure discovery learning?. *The American Psychologist*, *59*(1), 14–19. doi:10.1037/0003-066X.59.1.14 PMID:14736316

Mazzoni, G., & Vannucci, M. (2007). Hindsight bias, the misinformation effect, and false autobiographical memories. *Social Cognition*, *25*, 203–220. doi:10.1521/soco.2007.25.1.203

McCall, R. B. (1977). Challenges to a science of developmental psychology. *Child Development*, *48*(2), 333–344. doi:10.2307/1128626

Mehlenbacher, B., Miller, C. R., Covington, D., & Larsen, J. S. (2000). Active and interactive learning online: A comparison of Web-based and conventional writing classes. *IEEE Transactions on Professional Communication*, *43*(2), 166–184. doi:10.1109/47.843644

Mellers, B., Stone, E., Atanasov, P., Rohrbaugh, N., Metz, S. E., Ungar, L., & Tetlock, P. et al. (2015). The psychology of intelligence analysis: Drivers of prediction accuracy in world politics. *Journal of Experimental Psychology. Applied*, *21*(1), 1–14. doi:10.1037/xap0000040 PMID:25581088

Mellers, B., Stone, E., Murray, T., Minster, A., Rohrbaugh, N., & Bishop, M. (2015). Identifying and cultivating superforecasters as a method of improving probabilistic predictions. *Perspectives on Psychological Science*, *10*, 267-281. doi: 1745691615577794

Mood, T. A. (1995). *Distance education: An annotated bibliography*. Englewood, CO: Libraries Unlimited, Inc.

Moore, D. S. (1997). New pedagogy and new content: The case of statistics. *International Statistical Review*, *65*(2), 123–137. doi:10.1111/j.1751-5823.1997.tb00390.x

Moore, M. G., & Kearsley, G. (1996). *Distance education: A systems view. Wadsworth*. Belmont, CA: Wadsworth Publishing.

Morewedge, C. K., & Kahneman, D. (2010). Associative processes in intuitive judgment. *Trends in Cognitive Sciences*, *14*(10), 435–440. doi:10.1016/j.tics.2010.07.004 PMID:20696611

Mrazek, M. D., Franklin, M. S., Phillips, D. T., Baird, B., & Schooler, J. W. (2013). Mindfulness training improves working memory capacity and GRE performance while reducing mind wandering. *Psychological Science*, *24*(5), 776–781. doi:10.1177/0956797612459659 PMID:23538911

Mueller, P. A., & Oppenheimer, D. M. (2014). The pen is mightier than the keyboard advantages of longhand over laptop note taking. *Psychological Science*, *25*(6), 1159–1168. doi:10.1177/0956797614524581 PMID:24760141

Muraven, M., & Baumeister, R. F. (2000). Self-regulation and depletion of limited resources: Does self-control resemble a muscle?. *Psychological Bulletin*, *126*(2), 247–258. doi:10.1037/0033-2909.126.2.247 PMID:10748642

Muraven, M., Shmueli, D., & Burkley, E. (2006). Conserving self-control strength. *Journal of Personality and Social Psychology*, *91*(3), 524–537. doi:10.1037/0022-3514.91.3.524 PMID:16938035

Norton, B. (2016). Sanders' supporters cry foul over "Coingate": Controversy over coin tosses that made Clinton "winner" in Iowa. *Salon*. Retrieved from: http://www.salon.com/2016/02/02/critics_cry_foul_over_coingate_hillary_clinton_had_1 _6_chance_of_winning_6_coin_tosses_that_made_her_winner_in_iowa/

Oakes, M. (1986). *Statistical inference: A commentary for the social and behavioural sciences*. Chichester, UK: Wiley.

Paas, F., Tuovinen, J. E., Tabbers, H., & Van Gerven, P. W. (2003). Cognitive load measurement as a means to advance cognitive load theory. *Educational Psychologist*, *38*(1), 63–71. doi:10.1207/S15326985EP3801_8

Pajares, F. (2003). Self-efficacy beliefs, motivation, and achievement in writing: A review of the literature. *Reading & Writing Quarterly*, *19*(2), 139–158. doi:10.1080/10573560308222

Pashler, H. (1984). Processing stages in overlapping tasks: Evidence for a central bottleneck. *Journal of Experimental Psychology. Human Perception and Performance*, *10*(3), 358–377. doi:10.1037/0096-1523.10.3.358 PMID:6242412

Passolunghi, M. C., & Siegel, L. S. (2004). Working memory and access to numerical information in children with disability in mathematics. *Journal of Experimental Child Psychology*, *88*(4), 348–367. doi:10.1016/j.jecp.2004.04.002 PMID:15265681

Perkins, D. V., & Saris, R. N. (2001). A jigsaw classroom technique for undergraduate statistics courses. *Teaching of Psychology*, *28*(2), 111–113. doi:10.1207/S15328023TOP2802_09

Persson, J., Welsh, K. M., Jonides, J., & Reuter-Lorenz, P. A. (2007). Cognitive fatigue of executive processes: Interaction between interference resolution tasks. *Neuropsychologia*, *45*(7), 1571–1579. doi:10.1016/j.neuropsychologia.2006.12.007 PMID:17227678

Pfaff, T. J., & Weinberg, A. (2009). Do hands-on activities increase student understanding? A case study. *Journal of Statistics Education*, *17*, 1–34. Retrieved from https://ww2.amstat.org/publications/jse/v17n3/pfaff.pdf PMID:21603584

Pfannkuch, M., & Wild, C. (2004). Towards an understanding of statistical thinking. In D. Ben- Zvi & J.B. Garfield (Eds.), *The challenge of developing statistical literacy, reasoning and thinking* (pp. 17-46). Dordrecht, Netherlands: Kluwer Academic Publishers. doi:10.1007/1-4020-2278-6_2

Pfister, H. R., & Böhm, G. (2008). The multiplicity of emotions: A framework of emotional functions in decision making. *Judgment and Decision Making*, *3*, 5–17.

Piattelli-Palmarini, M. (1994). *Inevitable illusions. How mistakes of reason rule our minds*. New York, NY: Wiley.

Piolat, A., Olive, T., & Kellogg, R. T. (2005). Cognitive effort during note taking. *Applied Cognitive Psychology*, *19*(3), 291–312. doi:10.1002/acp.1086

Posner, M. I., & Snyder, C. R. (2004). Attention and cognitive control. In D. A. Balota & E. J. Marsh (Eds.), *Cognitive psychology: Key readings*, (pp. 205-223). New York, NY: Psychology Press.

Prince, M. (2004). Does active learning work? A review of the research. *The Journal of Engineering Education*, *93*(3), 223–231. doi:10.1002/j.2168-9830.2004.tb00809.x

Pyc, M. A., & Rawson, K. A. (2009). Testing the retrieval effort hypothesis: Does greater difficulty correctly recalling information lead to higher levels of memory?. *Journal of Memory and Language*, *60*(4), 437–447. doi:10.1016/j.jml.2009.01.004

Rickards, J. P., & Friedman, F. (1978). The encoding versus the external storage hypothesis in note taking. *Contemporary Educational Psychology*, *3*(2), 136–143. doi:10.1016/0361-476X(78)90020-6

Rowe, N. C. (2004). Cheating in online student assessment: Beyond plagiarism. *On-Line Journal of Distance Learning Administration, 7*. Retrieved from: http://www.westga.edu/~distance/ojdla/summer72/rowe72.html

Salmon, G. (2013). *E-tivities: The key to active online learning*. New York, NY: Routledge.

Sana, F., Weston, T., & Cepeda, N. J. (2013). Laptop multitasking hinders classroom learning for both users and nearby peers. *Computers & Education*, *62*, 24–31. doi:10.1016/j.compedu.2012.10.003

Schmeichel, B. J., Vohs, K. D., & Baumeister, R. F. (2003). Intellectual performance and ego depletion: Role of the self in logical reasoning and other information processing. *Journal of Personality and Social Psychology*, *85*(1), 33–46. doi:10.1037/0022-3514.85.1.33 PMID:12872883

Schochet, P. Z. (2009). Statistical power for regression discontinuity designs in education evaluations. *Journal of Educational and Behavioral Statistics*, *34*(2), 238–266. doi:10.3102/1076998609332748

Schunk, D. H. (1991). Self-efficacy and academic motivation. *Educational Psychologist*, *26*(3-4), 207–231. doi:10.1080/00461520.1991.9653133

Shea, P., Li, C. S., & Pickett, A. (2006). A study of teaching presence and student sense of learning community in fully online and web-enhanced college courses. *The Internet and Higher Education*, *9*(3), 175–190. doi:10.1016/j.iheduc.2006.06.005

Simon, H. A. (1955). A behavioral model of rational choice. *The Quarterly Journal of Economics*, *69*(1), 99–118. doi:10.2307/1884852

Simon, M., Houghton, S. M., & Aquino, K. (2000). Cognitive biases, risk perception, and venture formation: How individuals decide to start companies. *Journal of Business Venturing*, *15*(2), 113–134. doi:10.1016/S0883-9026(98)00003-2

Snee, R. D. (1993). What's missing in statistical education?. *The American Statistician*, *47*, 149–154. doi:10.1080/00031305.1993.10475964

Sommerville, J. A., Bernstein, D. M., & Meltzoff, A. N. (2013). Measuring beliefs in centimeters: Private knowledge biases preschoolers and adults representation of others beliefs. *Child Development*, *84*(6), 1846–1854. doi:10.1111/cdev.12110 PMID:23581849

Sotos, A. E. C., Vanhoof, S., Van den Noortgate, W., & Onghena, P. (2007). Students misconceptions of statistical inference: A review of the empirical evidence from research on statistics education. *Educational Research Review*, *2*(2), 98–113. doi:10.1016/j.edurev.2007.04.001

Stahl, G. (2006). *Group cognition: Computer support for building collaborative knowledge (acting with technology)*. Cambridge, MA: The MIT Press.

Steinkuehler, C. A. (2008). Cognition and literacy in massively multiplayer online games. In J. Coiro, M. Knobel, C. Lankshear, & D. Leu (Eds.), *Handbook of research on new literacies* (pp. 611–634). New York, NY: Routledge.

Strayer, D. L., & Drews, F. A. (2007). Cell phone induced driver distraction. *Current Directions in Psychological Science*, *16*(3), 128–131. doi:10.1111/j.1467-8721.2007.00489.x

Strayer, D. L., Drews, F. A., & Crouch, D. J. (2006). A comparison of the cell phone driver and the drunk driver. *Human Factors*, *48*(2), 381–391. doi:10.1518/001872006777724471 PMID:16884056

Sustein, C.R., Bobadilla-Suarez, S., Lazzaro, S.C., & Sharot, T. (2016). *How people update beliefs about climate change: Good news and bad news*. doi: 10.2139/ssrn.2821919

Swan, K. (2001). Virtual interaction: Design factors affecting student satisfaction and perceived learning in asynchronous online courses. *Distance Education*, *22*(2), 306–331. doi:10.1080/0158791010220208

Swan, K., & Shih, L. F. (2005). On the nature and development of social presence in online course discussions. *Journal of Asynchronous Learning Networks*, *9*, 115–136.

Tetlock, P. E., & Gardner, D. (2016). *Superforecasting: The art and science of prediction*. New York, NY: Broadway Books.

Tversky, A., & Kahneman, D. (1973a). Availability: A heuristic for judging frequency and probability. *Cognitive Psychology*, *5*(2), 207–232. doi:10.1016/0010-0285(73)90033-9

Tversky, A., & Kahneman, D. (1974). Heuristics and biases: Judgement under uncertainty. *Science*, *185*, 1124–1130. doi:10.1126/science.185.4157.1124 PMID:17835457

Tversky, A., Kahneman, D., & Moser, P. (1990). Judgment under uncertainty: Heuristics and biases. In P. K. Moser (Ed.), *Rationality in action: Contemporary approaches* (pp. 171–188). New York, NY: Cambridge University Press.

Vaughan, N., & Garrison, D. R. (2005). Creating cognitive presence in a blended faculty development community. *The Internet and Higher Education*, *8*(1), 1–12. doi:10.1016/j.iheduc.2004.11.001

Vie, S. (2008). Digital divide 2.0: Generation M and online social networking sites in the composition classroom. *Computers and Composition*, *25*(1), 9–23. doi:10.1016/j.compcom.2007.09.004

Vygotsky, L. S. (1986). *Thought and language (E. Hanfmann & G, Vakar, Trans.)*. Cambridge, MA: MIT Press.

Wang, S. K., & Reeves, T. C. (2007). The effects of a web-based learning environment on student motivation in a high school earth science course. *Educational Technology Research and Development*, *55*(2), 169–192. doi:10.1007/s11423-006-9016-3

Wang, X. T., Simons, F., & Brédart, S. (2001). Social cues and verbal framing in risky choice. *Journal of Behavioral Decision Making*, *14*, 1–15. doi:10.1002/1099-0771(200101)14:1<1::AID-BDM361>3.0.CO;2-N

Warner, C. B., & Meehan, A. M. (2001). Microsoft Excel™ as a tool for teaching basic statistics. *Teaching of Psychology*, *28*(4), 295–298. doi:10.1207/S15328023TOP2804_11

Weinstein, Y., McDermott, K. B., & Roediger, H. L. III. (2010). A comparison of study strategies for passages: Rereading, answering questions, and generating questions. *Journal of Experimental Psychology. Applied*, *16*(3), 308–316. doi:10.1037/a0020992 PMID:20853989

Weltman, D., & Whiteside, M. (2010). Comparing the effectiveness of traditional and active learning methods in business statistics: Convergence to the mean. *Journal of Statistics Education*, *18*, 1–13.

Wilson, K. M., & Swanson, H. L. (2001). Are mathematics disabilities due to a domain-general or a domain-specific working memory deficit?. *Journal of Learning Disabilities*, *34*(3), 237–248. doi:10.1177/002221940103400304 PMID:15499878

Woolley, T. W. (2013). Teaching outside the box: Business statistics in the context of big question of meaning and value. *Journal of the Academy of Business Education*, *13*, 97–106.

Wright, M., De Geus, E., Ando, J., Luciano, M., Posthuma, D., & Ono, N. et al.. (2001). Genetics of cognition. Outline of a collaborative twin study. *Twin Research*, *4*, 48-56. doi: Wild, C.J., & Pfannkuch, M. (1999). Statistical thinking in empirical enquiry. *International Statistical Review*, *67*, 223–248. doi:10.1111/j.1751-5823.1999.tb00442.x

ENDNOTE

1 While college students themselves may—or may not—be ethnically diverse at any given higher educational institution, it should be noted that statistics classes are often diverse in terms of degree focus and plan of study.

Chapter 4
A Cognitive Curriculum:
Improving Statistics Cognition Online

ABSTRACT

This chapter provides a discussion of developing a curriculum for a modern statistics course which aims to improve statistics cognition. We begin by examining micro-level curricular considerations, such as designing learning objectives and assessments which can allow transfer of cognitive processes. Then, we discuss the implications of macro-level curricular considerations, such as tracking, and the need to search for a mismatch between the learner and their environment. Collectively, we argue that such practices allow educators to develop a cognitive curriculum. We conclude the chapter with a discussion of how online learning environments inherently lend themselves to a cognitive curriculum and provide numerous benefits for the educator and learner.

The design of the physical environment itself is equally important as content and pedagogy. - J.D. Mills, 2015, p. 62

1. INTRODUCTION

So far, discussion has centered on statistics cognition as internal constructs researchers and educators can empirically measure (Chapter 2), and as a central mechanism by that scholars can shed light on student learning (Chapter

DOI: 10.4018/978-1-5225-2420-5.ch004

3). However, cognitions is not static, and may often depend on the situation (see Brown, Collins, Duguid, 1989; Kishner & Whitson, 1997). Indeed, a common assumption made by educational scholars is that learning itself is situated, for if we did not believe that formal settings enhanced education, we probably would not have schools.

In order to understand how cognition can be situated, we look toward the field of *curriculum* (see Tanner & Tanner, 1980). While the definition of curriculum has changed over time (Kelly, 2009), it may often be thought of as the culmination of an educational experience (Pinar, 1995). That is, how students learn knowledge or skills, and what they end up doing with this information. While deciding what curriculum to employ can be a major undertaking in of itself (see Gall, 1981), an array of resources have emerged to help educators select curriculum for disciplines, such as statistics (Bakker, 2004; Cheatham, 2000; Cobb, 1993; 2007), as well as modalities, such as online learning (Edelson, 2001; Edelson, Gordin, & Pea, 1999; Quintana et al., 2004).

Although relatively fewer studies have examined curriculum of online statistics courses (Summers, Waigandt, & Whittaker, 2005), it is imperative to consider how quality education may be independent of certain environmental features, depending on the student. As previously mentioned, while meta-analyses have found students in online courses are relatively equal in terms of their performance and affect (see Allen, Bourhis, Burrell, & Mabry, 2002; Allen, Mabry, Mattrey, Bourhis, Titsworth, & Burrell, 2004), results vary widely depending upon the quality of the curriculum (Zhao, Li, Yan, Lai, Tan, 2005).

Although more recent meta-analyses have suggested that technology may provide a slight advantage to students, relative to classes that only rely on face-to-face instruction (Means, Toyama, Murphy, Bakia, & Jones, 2009; Sitzmann, Kraiger, Stewart, & Wisher, 2006), this Chapter discusses the overarching framework for developing a *cognitive curriculum*, or rather, a curriculum that assesses current levels of cognitive ability, and implements materials that prompt the student to go beyond their current means. In a classical social-constructivist sense, the educator assesses the limits of a student's cognition, and targets instruction on what the student can do with assistance (Vygotsky, 1986).

While this Chapter integrates research on statistics curriculum, many comprehensive reviews of curriculum and resources for statistics education exist elsewhere (Cobb, 1993; 2007; Bryce, Gould, Notz, & Peck, 2001; Franklin et al., 2005). Whether developing assessments (Gal & Garfield, 1997), aligning statistics curriculum with related disciplines (Bargagliotti, 2012), or examining the role of technology in improving statistics education (Chance, Ben-Zvi, Garfield, & Medina, 2007), researchers have amassed a wealth of knowledge about designing statistics courses (Bradstreet, 1996; Dietz-Uhler, Fisher, & Han, 2007; Remsburg, Harris, & Batzli, 2014). Thus, instead of reiterating this material, our focus is on research exploring the ways online curriculum can be used to facilitate cognitive growth. This topic will be approached from two primary perspectives.

This Chapter begins by addressing micro-level curriculum considerations, focusing on curricular features that are meant to interact with statistics cognitions. We argue that by framing learning objectives on cognitions (2.1), and using cognitively appropriate assessments (2.2), transfer of cognition - or the propensity to see the differential use of a cognitive skill throughout a variety of situations - may be achievable (2.3). We then go on to discuss the macro-level curricular considerations, focusing on how educators identify students who are below or beyond the curricular requirements for a particular course (3.1). We go on to discuss how curricular mismatches are typically resolved (3.2), and conclude this sub-section with presenting new paradigms on solving curricular mismatches (3.3). We then shift our discussion toward the benefits of online education (4.1), outlining ways common obstacles can be overcome (4.2). We conclude with a summary of online cognitive curriculum (5).

2. MICRO-LEVEL CURRICULUM FOR STATISTICS COGNITION

To examine the features of micro-level curriculum, it is necessary to understand how curriculum can be structured to influence cognitions. To achieve this, we maintain that the educator must employ the scientific method and develop a *design experiment* (Brown, 1992) by which carefully crafted learning objectives are repeatedly assessed so that materials may be refined, and personal cognitive growth may be gauged. To do this, the educator must first ensure that their learning objectives are cognitive in nature.

2.1. Cognitive Learning Objectives

Cognitive learning objectives are those framed on cognitive - as opposed to behavioral - tasks (Mayer, 2004). Although behavioral tasks, such as computing an ANOVA on M&M color (Froelich & Stephenson 2013; Schwartz, 2013), could result in a fun learning experience, enhanced learning does not occur just by virtue of using an activity (Pfaff & Weinberg, 2009). However, with that said, it is necessary to incorporate some behavioral components into learning objectives, as cognition may not easily be parsed within a classroom environment. While there should be a strong emphasis to developing cognitive objectives, what is most important is that each objective is designed to promote statistics literacy, reasoning, and thinking (Chapter 2).

Creating curriculum from scratch can often seem daunting, and takes a lot of work to verify the quality of educational content. However, professors in higher education may often lack the time and resources needed to develop stringent curriculum. Luckily scholarly work (see Ben-Zvi & Garfield, 2004; Tintle, VanderStoep, Homles, Quisenberry, & Swanson, 2011) and professional organizations (Aliaga et al., 2005) have laid out many options for educators. For example, guidelines established by the American Statistical Association (Franklin et al., 2005; Metz, 2010) propose twenty-two goals (Table 1) and framed on key actions, such as "*understanding why*" or "*recognizing*". Importantly, using the previously outline framework (Chapter 2), these key actions allow the educator to indicate the types of cognitions that their objectives are designed to develop.

For instance, consider the goal, "students should *recognize* terms such as 'normal', 'random', 'correlation'." This goal asserts that students must know the terms "normal", "random" and "correlation" as they apply to statistics. As an educator, one must consider how to embed this goal into their curriculum. To do this, the educator must first consider how they will convey what is important to each student. Whether by direct lecture or demonstration, the students must be prompted to encode specific information. Then, the educator must decide their data collection plan, building a way to measure every learning objective (Brown, 1992). Largely, this process involves critically thinking about how students will achieve these goals, and what resources will be provided to help them succeed.

Importantly, a unique assessment is not necessary for every learning goal. Much like the learning objectives themselves are connected (see Figure 2 of Chapter 2), course objectives may often include multiple cognitive processes.

Table 1. Educational guidelines arranged according to the neural education model

	Literacy	Reasoning	Thinking
Students Should Believe and Understand Why			
Data beat anecdotes	X	-> X	
Variability is natural, predicTable, quantifiable	X	-> X	
Random sampling allows generalization	X	-> X	
Random assignment allows causal claims	X	-> X	
Association is not causation	X	-> X	
Statistical significance does not imply importance	X	-> X	
Non-significant effects may not mean no relationship	X	-> X	
Students Should Recognize			
Common sources of bias in surveys and experiments	X		
Determine the population which results allow inference	X	-> X	
Determine when causal inference is warranted	X	-> X	
Terms such as "normal", "random", "correlation"	X		
Students Should Understand			
How to obtain/generate data	X	-> X	-> X
How to graph data as first step	X	-> X	-> X
How to interpret data	X	-> X	-> X
How to make appropriate use of statistical inference	X	-> X	-> X
How to communicate	X	-> X	-> X
Students Should Understand Statistical Inference			
Sampling distribution and how it applies to sampling	X	-> X	
Concept of statistical significance and p-values	X	-> X	
Concept of confidence intervals and interpretation	X	-> X	
Students Should Know			
How to interpret results in context	X	-> X	-> X
How to critique a variety of media	X	-> X	-> X
When to call for help from a statistician	X	-> X	-> X

Note: The GAISE guidelines suggest learning goals, each corresponding with one or more statistics cognition. Once a goal requires a literacy outcome, it can be built upon (see "->" above), to include reasoning, and thinking cognitions.

For instance, consider the learning objective "students should understand how to interpret data." This skill will assume that students have adequate literacy, or be inherently linked to separate learning objectives that explicitly outline the literacy skills needed to interpret data. Further, to interpret data, students must also be able to exercise some degree of reasoning. That is, students will need to know how to recognize the features of the data, and be able to reason what the data could be used for.

On of the barriers to developing such a curriculum is that, until now (Chapter 2), there was no model directly linking cognitive processes with educational objectives. While field fragmentation could be problematic for the future of statistics education (Beyth-marom, Fidler, Cumming, 2008), the divide among scholarly pursuits is more structural than it is related to the content of their studies. Regardless as to whether one is a neuroscientist, academic researcher, or educator, the process of collecting data to support claims remains constant.

In fact, what differs is the focal point, each with a different burden of proof in assessing relevant constructs (Table 2). For example, neuroscientists may need to prove that encoding exists as a distinct construct, whereas statistics education scholars focus more on the desired outcome of appropriate encoding, namely literacy of statistical constructs. Educators take this a step further by not directly measuring whether encoding resulted in literacy, but rather prompting students to demonstrate their ability to encode via retrieval.

If cognitive processes, such as encoding, planning, and problem solving, can indeed be linked to statistics cognitions, then a great deal of consideration needs to be given toward how statistics cognitions are measured. When conducting any scientific experiment, the way that variables are defined, and ultimately measured, can have a drastic impact on how findings can be interpreted. For example, consider the literature summarized in Chapter 1, that suggests online learning can be equivalent to face-to-face modes of instruction (Bernard et al., 2004). Although such a finding may be a relief to those who provide online education, it is hardly informative to the scholar,

Table 2. Focus of statistics education research by type of professional

Neuroscientists	Neuroscientists are interested in the way the brain develops and functions. Much research in neuroscience is dedicated to understanding how localized the human brain is, and whether or not distinct cognitive functions exist. Currently research (see Anderson, Pyke, & Fincham, 2016) suggests processes for encoding, planning, and problem solving.
SE Scholars	Scholars studying statistics education are concerned with the way specific cognitions can interact with the way people understand or make use of statistical information. Scholars suggest (see Ben-Zvi & Garfield, 2004) domain specific cognitions for literacy, reasoning, and thinking.
Educators	Statistics educators should primarily be interested in obtaining cognitive learning objectives (see Franklin et al., 2005). Ideally, and educator is helping a student develop their literacy, reasoning, and thinking skills, and must collect evidence providing support for each cognition.

especially when considering how the sheer diversity in program quality makes it extremely difficult to determine if differences really exist between modalities (Zhao et al., 2005). Similarly, if neuroscientists, statistics education scholars, and educators all use different definitions and methods of measurement, the reliability and validity of establishing a research program based on these interdisciplinary studies is greatly reduced. Much like the scientist needs to take great care in developing appropriate instruments to determine whether an effect is present, the educator must carefully construct assessments that are able to consistently assess cognitive growth.

2.2. The Role of Assessments in Cognitive Curriculum

While neuroscientists and scholars studying statistics education may have a variety of tools at their disposal, educators may find it difficult to directly measure cognition in the classroom. For instance, as previously mentioned, educators may not be assessing encoding as much as they are retrieval. While such processes may share commonalities, the two may not always coincide, and an educator may be left trying to determine whether a student has not been studying (failure to encode) or whether the student experiences test anxiety that creates a cognitive load which may interrupt retrieval (Chapter 5). If assessments are an educators evidence (Brown, 1992), then assessments must be crafted to provide information about student's cognitive development.

While researchers are trained with a high degree of scientific rigor, educators in higher education may not be formally trained as educators. As a result, they may have never been required to take an assessment course, and may not understand the need to approach their assessments with the same rigor as developing a psychometrically sound scale. However, imagine for a moment that you are teaching statistics and want evidence to prove to your colleagues or superior that your students "understand how to communicate statistical information". One option is to design a simple true or false statement. Ask students to identify which of the two statements is the correct way to report a p-value. However, does this proof convince you? If this were a dire scientific finding, would your colleagues all nod in agreement that you have in fact met this objective?

Perhaps a better way of demonstrating this is to have students take statistical information and communicate it themselves. Yet, a concern of ecological validity may arise (Shadish, Cook, & Campbell, 2002), when translating this task to an assessment. After all, how often are students explicitly asked "state your decision criteria" and "state whether or not you rejected the null hypothesis" when conducting research in their future careers? Such artificial ways to assess students do not adequately measure cognitive gains, and could even may hinder learning (see 2.3). Communication is about understanding information and conveying the important elements of it to others. As a result, it may be even better to have students create their own empirical research question, collect and analyze data, only to have each to present it in front of the class (Chance, 1997; Gal & Garfield, 1997; Garfield & Gal, 1999). At least, evidence from this form of *authentic* assessment may be more convincing when used to evaluate how well a student communicates statistical information.

However, even with the best assessments, gathering proof for educational objectives requires triangulation from numerous sources (Rothbauer, 2008). For example, the GAISE (Franklin et al., 2005) objectives are not likely to be assessed on a single assignment, but rather throughout the span of a course. As a result, educators may often find that they need to develop sub-goals under each objective to specify the nature of the learning experiences. By crafting detailed learning objectives, the educator helps guide student encoding by dictating the relative importance of tasks (Bonner, 1999; Gronlund, 1985). Additionally, by crafting goals that make use of multiple sources, the educator, much like a psychometrician, can examine the value each task has in helping students reach that overarching learning objective. Yet, even with a stringent curricular design framed on cognitive objectives, there are several obstacles educators may face in assessing cognition.

Perhaps one of the more apparent obstacles faced is whether educational methods are valid. Although neuroscientists can use an fMRI to assess the activity of encoding, planning, and general problem (Anderson et al., 2016), educators cannot easily assess what a student is doing when they are working on a problem. Thus, in order to obtain evidence for constructs such as "statistical thinking", educators need to be more clever, and specific. In other words, it is not enough that an educator assume that a student's correct or incorrect response is a true gauge of their problem solving cognition. Such an error could be due to a variety of factors (Chapter 3). Instead, educators must observe application, and have students show their work to provide evidence that they are using their problem solving skills adequately. While problem-solving may be the baseline for neuroscientists, it is often the more complex dimension to assess in the classroom.

Another obstacle for educators is that many curricular tasks in statistics education are comprised of a series of cognitions. For example, imagine that you want to teach a student how to test a null hypothesis (Figure 1).

Figure 1. An outline of a five-step process to teach null hypothesis testing by cognitive goal

Null hypothesis steps	Literacy (Remember)	Reasoning (Understanding)	Thinking (Applying)
1) State the null and alternative hypotheses	**Define** and **describe** the null and alternative hypotheses.		
2) Select the appropriate test statistic	**Recall** possible test statistics, **identify** appropriate test statistic	**Interpret** information stated in the problem to help determine appropriate test statistic	
3) Establish decision rules which one would use to interpret their results as statistically significant	**Recall** possible decision criteria, **identify** appropriate decision criteria, **define** decision criteria	**Explain** why decision criteria was selected	
4) Compute the appropriate test statistic	**Recall** appropriate mathematical formula	**Interpret** test relevant information (i.e., uneven *n*).	Use the formula, **solve** your research question
5) Make a decision	**Recall** rules for interpreting p-values	**Summarize** results, **interpret** data, **infer** decision, **explain** data and decision	Use data to generalize

For simplicity sake, an educator may develop a five step null-hypothesis process, that prompts the student to follow a series of steps. However, each step involves a great deal of literacy of different content, with sub-processes related to reasoning and thinking being applied. Although asking students to show work is an ideal method to identify and troubleshoot errors student make, students may not always show their work, and even when they do, their line of reasoning may not be entirely clear.

While task categorization, such as that outlined in Figure 1, may seem tedious, it serves several important functions for a cognitive curriculum. First, it provides an educator with a bird's eye view of their curriculum. By categorizing tasks, educators state how much of their curriculum is designed to individually promote statistical literacy, reasoning, and thinking skills. Further, by make objectives specific, educators can assess how much each cognition grows, and identify patterns in learning or errors (Chapter 2). For instance, imagine that a student does well on homework, but tends to do poorly on exams. If the student does not generally feel anxious when taking exams (Chapter 5), it may suggest that the student is not applying the appropriate study strategy.

In a review of study strategies, researchers identified re-reading and underlining as having a lower utility relative to strategies that involve elaborate rehearsal (Dunlosky, Rawson, Marsh, Nathan, & Willingham, 2013). However, regardless of this finding, students commonly use low utility strategies - such as re-reading one's text - thinking it will lead to success (Karpicke, Butler, & Roediger, 2009). In a sense, there may not be anything wrong with re-reading, as it may help students clarify concepts and reinforce exposure to material. However, the success of re-reading largely depends on the type of tasks required of students as part of the curriculum. If tests, such as those in statistics, focus more on critical thinking and problem solving, re-reading notes is not likely to be effective. Because the utility of study strategies depends on the nature of the task (Weinstein, McDermott, & Roediger, 2010), educators must carefully ensure there is a relative match between objectives, tasks, and assessments.

Yet categorizing curriculum serves more than just allowing the educator to help an individual students overcome learning difficulties. Such information is also able to inform the educator about their success in developing student cognition. For example, an educator could look at various test problems over the course of the term that taps into statistical literacy, reasoning, and thinking. If the educator sees a class wide linear trajectory denoting a positive relationship among performance on measures of literacy, reasoning, and thinking, over the course of the term, the educator has found that they have helped successfully develop student statistics cognition (see Figure 2: example A). If, on the other hand, the educator notices a flat line near the top or bottom of the grade distribution, it may suggest that the curriculum did not present the right amount of challenge, or that the assessments did not adequately capture cognitive change over time (see Figure 2: example B). Such a finding would prompt an educator to review their method, and consider adjustments to their curriculum so that they can ensure that statistics cognition is being developed.

By taking care to categorize cognitions at the objective, activity, and assessment level, educators are ensuring a consistency of methods throughout their curriculum. Such a process is time consuming, but will be of great value in establishing how cognitions develop, and what the educator can do to ensure their course is efficiently and effectively enhancing cognitive development. Developing curriculum can be difficult, and just striving for alignment is not sufficient. Evidence must be collected to ensure curriculum inputs match their outputs. However, in doing so the educator must take care to avoid creating an artificial environment that takes a reductionist approach

Figure 2. Inferring from cognitive assessment patterns

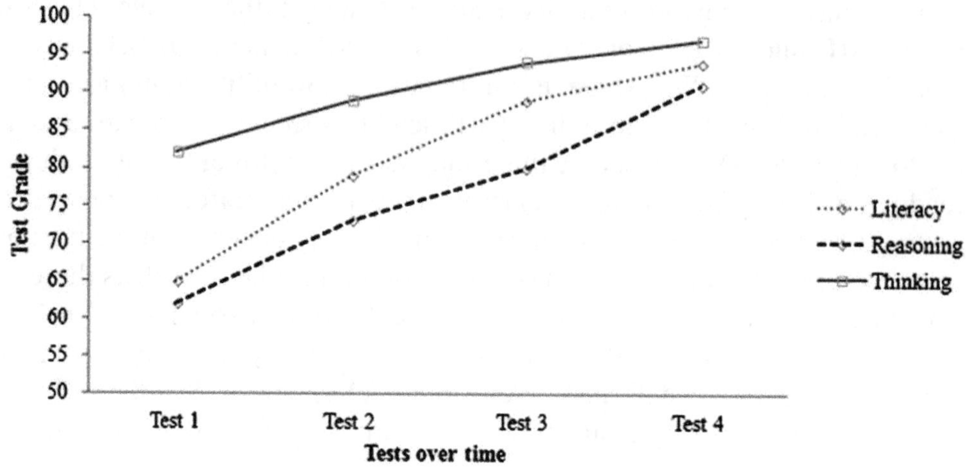

EXAMPLE A

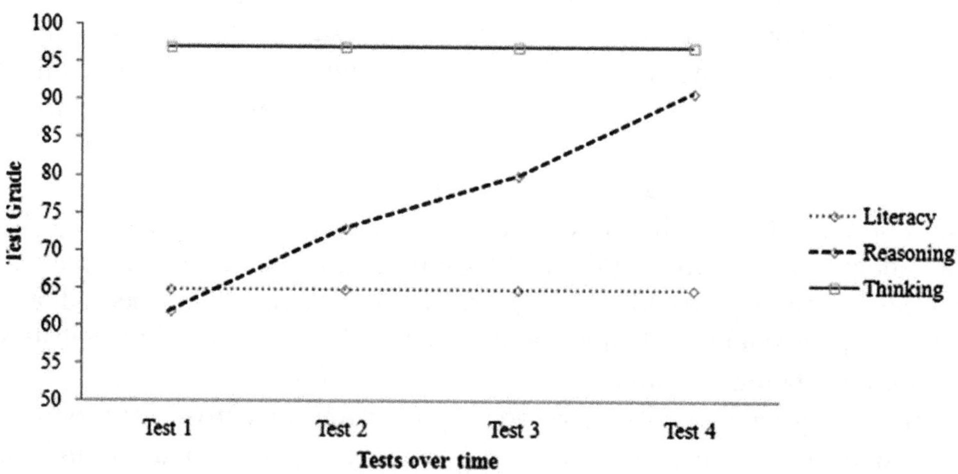

EXAMPLE B

to skill development. Statistics educators must strive for developing long-term skills which can be adapted to daily life.

2.3. Transfer of Cognition

In addition to developing a well aligned curriculum, educators concerned with establishing lifelong goals must assess and establish a student's ability to use their statistics cognitions in a variety of capacities. This process, referred to as the *transfer of cognition* depends on the use of authentic assessments. Although calls for revamping statistics curriculum throughout the 1990s largely asserted that statistics education must prepare students to deal with statistics in the real world (Friel & Corwin, 1990; Rouncefield, 1994; Scheaffer & Jaccobbe, 2014), there is a need ensure that the cognitive skills students are equipped with are flexible. While educators may hope that the tools students are equipped with in the classroom transfer to their daily lives, achieving transfer is exceedingly difficult (Lohman, 1993).

Although there is a limitation on what educators have control over, designing the learning environment to prompt students to use skills in different ways is central to transferring cognition. For instance children who learned mathematic operations by selling products on the streets of Brazil had a difficult time using these same skills on formal mathematics exams (Carraher, Carraher, & Schliemann, 1985). While there was no doubt that these students understand certain mathematic operations, it was difficult for students to apply these skills to a formal setting. Similarly, a student in a University level statistics classroom could walk out of the classroom, not realizing that the skills they are developing in this class can be used in a variety of settings. Such a finding is consistent with other results interpreted to suggest that one's memory can be influenced by a variety of environmental (Godden & Baddeley, 1975) or affective (Weingartner, Miller, & Murphy, 1977) features.

If the way material is organized can have a drastic impact on one's ability to learn, educators must consider how these skills will be used in daily life and plan activities that mimic the authentic use of these skills. However, in addition to mimicking environmental features, transfer may have to do with the order material is presented. For example, consider a statistics class that arranges content by topics, beginning with z-tests, moving into sampling theory, t-tests, ANOVA, and then (if enough time) correlation and regression.

As someone who has taken a variety of statistics courses, this progression is quite common in social science course offerings. However, such a presentation lacks a narrative arch, and may not explicitly demonstrate to students how these tools are connect.

Instead of arranging statistics curriculum by significance test, consider a statistics course that is framed on how and why people use statistics in science. Largely, the positivist approach means that scientists attempt to predict and describe events, thus, it may be fair to conceptualize a statistics class framed on describing and predicting events. Such a course could begin by centering on descriptive statistics (central tendency and correlation), and how these inform the learner with information that can be used to make an educated guess. Then, students can learn how to take data beyond description by building predictive models (regression). From here, the course could center on how to use statistical tools to inform one's predictive model (i.e., how t-tests, and ANOVA are used within the context of regression). By prompting students to ask themselves why they are carrying out specific operations, and to evaluate the quality of their statistical arguments, elusive and often neglected concepts such as confidence intervals and effect sizes (Cumming, 2012), become readily apparent as necessary indicators.

3. MACRO-LEVEL CURRICULUM FOR STATISTICS COGNITION

Learning objectives, assessments and activities are very much the building blocks of the curriculum. When selected and aligned with care, they may enhance cognitive outcomes. However, in addition to examining these building blocks, it is necessary to review curriculum from a macro-perspective. That is, it is necessary to examine how curriculum as a whole interacts with student learning. While statistics education scholars have long recognized the diverse needs of students in statistics (Wilks, 1951; Wishart, 1939), these discussions centered on the diversity of each academic discipline, as opposed to cognitive diversity within a student.

To appreciate why cognitive diversity matters, consider a hypothetical situation where an educator gives a pre-test to their class on the first day to establish a baseline of what students know, and where they need assistance.

Upon review, the educator may determine that students have varying educational needs (Tomlinson, 2003). In reviewing these needs, the educator identifies three unique groups of students. Perhaps the educator determines that while most students would be adequately challenge by the developed curriculum and demonstrate cognitive growth on classroom assessments, other groups would may need curricular adjustments. Perhaps one group of students demonstrates an advanced understanding of statistics, and as a result, could end up being bored by the curriculum. Conversely, another group may have demonstrated that they are behind the class average, and may need remediation to keep up with the pace of the course.

For an educator, diverse student ability is not only a potential problem for structural features such as content or pace, but it may also be a potential problem for interpreting assessments. Regardless of a student's ability, the objective of a statistics course is to enhance cognitive processes. As such, assessments in a cognitive classroom are required to measure cognitive development. This means that instead of comparing students to one another, they must be compared to themselves, showing progress by improving their understanding on assessments. However, because assessments are linked to curriculum, educators risk by giving inappropriate assessments when curriculum does not match student ability. In other words, assessments would not be reflective of growth, as students with advanced statistics cognitions would demonstrate a ceiling effect over time, whereas those who may need remediation would demonstrate a floor effect (Hessling & Schmidt, 2003).

Although such a dilemma is often not one an educator will find themselves in (some educators may lack direct knowledge of a student's educational capacity in such a clear-cut manner), it illustrates the complications that can arise from drastically differing zones of proximal development (Vygotsky, 1986). Although the previously outlined example is explicitly constructed to illustrate complications that arise when curriculum does not match student's ability, it does not undermine the importance of assessing curricular matching based on student cognitions.

Although curriculum mismatches can occur, they are not guaranteed to occur in every classroom. For example, depending on one's institution, an educator may have a set of students who are cognitively homogenous, thus negating any need to modify curriculum. Yet, when heterogeneity among students arises, there are a variety of ways educators have attempted to adjust for student differences, as well as new paradigms on the horizon that points to alternative routes for addressing curricular mismatches.

3.1. Resolving Curricular Mismatches

When using a cognitive curriculum, curricular mismatches can only be said to occur when a student has stunted or enhanced cognitions that do not appropriately align with planned curriculum. Exactly what degree constitutes a mismatch will likely depend on structural features, such as time and ability for educators to adapt curriculum, but more important should depend on the severity of the division in cognition.

Although separating students by ability may seem archaic, we note that prior divisions of ability grouping may have primarily been studied in classroom environments that focus on behavioral objectives (see Allan, 1991; Kulik & Kulik, 1982; 1984). Yet, when framing division on cognition, selection of student groups is based demonstrated difficulty or ease in using statistics cognitions. In other words, students who are unable to develop their statistics cognitions based on the current curriculum, as well as those who have already mastered their curriculum, may be suited for curricular adjustments.

To be clear, this is not to suggest that students with different general cognitions will automatically need curricular adjustments. As previously noted (Chapter 2), although the neural education (NE) model of statistics education aligns general level cognitions - such as problem solving ability - with domain specific cognitions - such as statistics thinking - there is no guarantee that those with different general cognitions will need different statistics curriculum. A student could have a great capacity for encoding information, but without the appropriate motivation (Dai & Sternberg, 2004; Shunk, Pintrich, & Meece, 2008), general cognitions may never be honed into their appropriate statistics cognitions. General cognitive ability may not necessarily be reflected in one's grades, and without the effort and persistence needed to undergo deliberate practice, cognitive expertise cannot be developed (Ericsson, Krampe, Tesch-Römer, 1993).

Considerations aside, schools have attempted to foster cognitive excellence and adjust for curricular mismatching by ability grouping in two distinct ways. The first involves ability grouping within a classroom, such as an educator who decides to put students in groups based on their ability, within the same classroom (Slavin, 1987). The second involves more of a structural approach, whereby administration constructs difference tracks, that students are assigned to based on their ability (Gamoran, 1992; Gamoran & Mare, 1989). While the promising results of these traditional solutions will be discussed, important nuances presented by using a cognitive curriculum suggest potential problems with these approaches.

3.1.1. Within-Classroom Grouping

One way to resolve a curricular mismatch is for an educator to sort students into groups based on their ability (Slavin, 1990). While it may seem difficult to gauge every student's cognitive ability, these groups are often informal, changing from subject-to-subject (Gamoran, 1992), and should not be considered part of a student's educational achievement. Further, groupings can be highly situational. For instance, an educator may assess student knowledge prior to the start of class, and then organize students based on whether or not they have mastered material from the previous class. Yet, over the course of the term, students may be moved among groups, as their cognitive progress can vary within the student (i.e., greater literacy relative to reasoning), and fluctuate over time.

Although the logic of grouping seems apparent, a question of whether or not grouping produces a difference in terms of educational outcomes exists. In a meta-analysis examining 52 studies, researchers found that students identified as gifted received a clear significant benefit for grouping, whereas average and below average students showed no benefit (Kulik & Kulik, 1982). Additionally, a second meta-analysis consisting of 145 studies found positive achievement effect for students assigned to small homogenous groups (Lou et al., 1996). While it seems that there is converging evidence to suggest that within-classroom grouping can improve student achievement, an educator must consider *how* and *why* within-classroom grouping is used.

For example, if an educator decides to group students based on their abilities, it may be clear to the students within the classroom which is the "top" and which is the "bottom" group (Slavin, 1992). Such expectations can form self-fulfilling prophecies (Eder, 1981), that could prompt students to endorse the label of the group they were assigned to define their success. Indeed, educational psychologists have long maintained that such expectations can play a central role in student outcomes (Rosenthal & Jacobson, 1968; Rubovits & Maehr, 1971).

Further, these studies frame within-classroom grouping as a cautionary tale, for educators must be careful to base within-classroom groups on classroom data, as opposed to a personal bias, or maladaptive benefit. For instance, one could imagine a situation that an educator groups students not based on ability, but rather, how the students are known to work together. An educator, seeking to save themselves a future headache, may be tempted to group students based on a symbiotic relationship. For example, imagine a student who prefers to work alone and typically dominates the groups they are assigned to work in. Then, imagine another student who shrugs off their responsibility, and refuses to contribute toward the collective effort. While this group may "perform" well, it should be noted the reasons for the achievement was maladaptive in that students did not make equal educational advances. As such, grouping in a cognitive curriculum should not be defined by ability, but rather growth trajectories.

For within-classroom ability grouping to be successful, it must offer real adjustments to curriculum. This is an effortful process (Lou et al., 1996; Tieso, 2003), that requires careful planning to successfully execute and monitor. Ultimately, the educator should first prove that grouping is necessary for their classroom by showing that they have students who demonstrate a consistently different cognitive ability, but also that methods involving grouping were superior in advancing cognitive growth. The purpose of within-classroom grouping is not to make an educators life easier, although after it is established, it could make classroom learning efficient (Lou et al., 1996; Saleh, Lazonder, De Jong, 2005; Sørensen & Hallinan, 1986). The purpose of ability grouping is to ensure that students are given the appropriate curriculum to enhance their domain specific cognitions.

3.1.2. Between Classroom Grouping (Tracking)

While it is possible for an educator to break their classroom into groups, it is also possible for administration to recognize the need for school wide grouping, designed to provide students with specific curriculum suited for their distinct educational needs (see Betts & Shkolnik, 2000; Epple, Newlon, & Romano, 2002). This process, commonly referred to as *tracking* (Oakes, 1987), serves a similar function to within grouping, but relies on a different approach (Rogers, 2002).

Primarily, between-classroom grouping allows educators to tailor entire lessons to groups of students (Ansalone, 2003), that often results in positive gains for gifted students (Rogers, 1991; 1993). Further, if within-classroom ability grouping risks creating a divide among students whereby stigma is attached to the low performing groups (Slavin, 1991), within-classroom ability grouping could negate the impact of such a stigma, at least at the classroom level.

However, there may be unintended costs associated with within class grouping. For example, educators who teach homogenous classes are likely to use fewer educational strategies to teach their students (Boaler, Wiliam, & Brown, 2000), which may have negative impacts on student enjoyment, leading to disengagement of course content (Chapter 5). Additionally, tracking has been shown that these tracks are often not being as homogenous as they could be (Hallinan, 1994), and worse yet, may be negatively biased toward students from disadvantaged backgrounds (Oakes, 1985; 1987). Indeed, low tracks tend to be comprised of students from low socioeconomic status households (Gamoran, 1987; Oakes, 1992), and could have a compounding negative effect on future educational attainment (Hanushek & Wößmann, 2006).

With a cognitive curriculum, it may make less sense to approach adjusting curriculum by using tracking. After all, cognitive development, from the educators' perspective, is a real time process to be managed over the course of a term. The fact is, not enough research exists to make specific predictions about how much statistics cognitions differ among students, and whether these differences are relatively stable. Even examining the predictive ability of general cognitive capacity is somewhat unclear. For instance, working memory capacity was somewhat related ($r = 0.30$) to ability in solving chemistry problems (Opdenacker et al., 1990), but the range of diversity in working memory capacity could vary by institution. That is, some colleges may find themselves having a large portion of high working memory capacity students, and as with between grouping, if a wide enough division does not exist among students, why implement structural changes to organizing classrooms?

3.2. A New Paradigm in Solving Curricular Mismatches

One of the problems with traditional ability grouping is that it generalizes curriculum at predefined levels that may or may not be appropriate for every student's need. These curricular adjustments are shortcuts, that categorize students in groups based on their relative abilities. However, much like the statistician who acknowledges that dichotomizing of continuous data results in a reduction of information, power, and inflates type II error (see Altman & Royston, 2006; Dawson & Weiss, 2012; Streiner, 2002), similar concerns could be present for design experiments (Brown, 1992).

For instance, consider splitting students into two groups based on their grade in the first half of a statistics course. If these students were grouped into high (100-85) and low categories (84 or <), then the educator is, in effect, suggesting there is no difference between a student with a 100 and an 85, nor would there be any difference between a student who earned an 84 and a 0. By collapsing across students in a general categorical fashion, the educator is losing some explanatory information regarding the student's learning, and could artificially create what looks like significant educational improve among groups, despite some students within these groups making no real progress.

Indeed, one of the largest concerns with ability grouping is that no one knows just how many groups to create. For example, research has found that the more tracks a school creates, the more homogenous grouping exists within each track (Hallinan, 1994). However, the reason why students are grouped together in a classroom is for efficiency, and it would not be practical to have hundreds of curriculum tracks within a school anymore than it would for an educator to create single person groups in a classroom.

Although whether or not ability grouping is seen as a viable option will depend on the instructor, a question emerges as to whose responsibility it is to deliver improved instruction to students. Although scholars have worked on drafting curricular resources for educators (see Cobb, 1993; Gal & Garfield, 1997), enrichment opportunities should also be capitalized on in the classroom to encourage students to apply their new skills to important problems that matter.

3.2.1. Statistical Enrichment

Student enrichment involves supplementing a student's educational experience. While curriculum offered by educational institutions is designed with great care to meet the needs of students (Tanner & Tanner, 1980), there may be some opportunities that schools are not well suited to provide. Schools are a limited resource, and are not able to provide every student with optimized individual curriculum, nor should they be burdened with such a task.

While it makes intuitive sense that enrichment activities would have a positive effect on student outcomes (Renzulli & Reis, 1995); meta-analytic results have shown mixed findings, with some finding that there is limited to no impact of extra-curricular activities on student outcomes (Shulruf, 2010), while others noting a positive influence on student outcomes (Kim & Bastedo, 2016). Yet, a lack of consistency in findings of enrichment activities does not imply that this approach is not viable, just that the mechanisms used to achieve cognitive development are highly complex. Depending on the structure of the activity, enrichment may enhance, have no effect on, or harm learning.

To clarify these possibilities Seow and Pan (2014) reviewed research on enrichment activities, identifying three distinct (development, threshold, and zero-sum) enrichment approaches that could have differential effects on learning. Specifically, the *development* framework (Fejgin, 1994) is based on student growth, aiming to highlight the strengths and weaknesses of a student in a particular domain, so that students may improve specific skills. Another theoretical perspective is the *threshold* framework, which argues that enrichment activities can improve academic outcomes up to a certain point (Fredricks & Eccles, 2010; Marsh, 1992), suggesting that low and moderate levels of enrichment positively impact student outcomes, but too much enrichment backfires. Still others have proposed a *zero-sum* framework (Coleman, 1961) whereby student outcomes were negatively linked to participation in enrichment, arguably because students focused more on enrichment than their academics.

Importantly, enrichment activities may largely be what students make of them. This motivational component is both a strength and weakness (see Dai & Chen, 2014). It may be a strength as it acknowledges the critical importance of student effort and motivation in learning; however, it creates unsystematic variance when examining educational outcomes. Motivation relies on resources, and just like directing self-regulatory behaviors that can become depleted (see Baumeister & Vohs, 2007; Hagger, Wood, Stiff, & Chatzisarantis, 2010; Muraven & Baumeister, 2000) motivation can change over time, "muddying" the ability to measure a direct benefit of enrichment on student outcomes.

Regardless of their benefit on performance, enrichment activities can have a positive influence on student engagement (Martin, 2008; Martin et al., 2013; Shernoff & Vandell, 2007), which may ultimately improve student perceptions of statistics education, and could promote long term retention of relevant skills. Professional organizations have the power to develop statistics enrichment programs that apply a developmental framework, and more scholarly work needs to focus on the efficacy and availability of such programs. Enrichment is but one option, but it uniquely allows students to apply practical skills and learn their utility, that may be artificial in a classroom environment that uses arbitrary cut-offs to group students. However, while grouping as we have previously discussed, may not be ideal, the principles that underlie grouping are relevant when considering how statistics curriculum can be differentiated.

3.2.2. Statistical Differentiation

Differentiation consists of altering a student's curriculum at some capacity in order to provide them with the appropriate academic challenge (Matthews & Foster, 2005). Largely, those in favor of differentiation argue that it allows educators to respond to the needs of gifted students on an individual basis (Dai & Chen, 2014). However, regardless of the source, a wealth of information exists suggesting that forms of differentiation, such as acceleration, are largely effective (Merrotsy, 2008).

For example, Park, Lubinski and Benbow (2013) analyzed data from a 40-year longitudinal study examining long term STEM outcomes for students who were accelerated at least one grade. Analyzing 363 accelerated students on 657 matched controls, Park and colleagues found that students who skipped a grade were more likely to pursue an advanced STEM degree, earned their degree and first publication earlier, and were more likely to produce work that would be highly cited, relative to matched non-grade skipping counterparts. Indeed, a variety of research has long demonstrated a slew of positive academic and social outcomes for students who experience a form of differentiation (Kulik & Kulik, 1984; Steenbergen-Hu & Moon, 2011; Swiatek & Benbow, 1991).

However, differentiation in its most extreme form, would encompass student cognitive needs, and attend to each student on an individual basis. While it is tempting to suggest that some students may be more (or less) cognitively similar to one another, there are a variety of important educational factors, such as motivation, that could explain how initially homogenous groups could become less homogenous over time. Because the primary goal is to realign curriculum to suit student needs, the techniques for differentiation are largely built upon the same framework. That is, by adjusting curriculum for students, student outcomes can be improved.

While the exact method of differentiation can be viewed as anything from early admission to kindergarten (Sankar-DeLeeuw, 1999) to credit by examination (Rogers, 2007), to curriculum compacting (Renzulli, Smith, & Reis, 1982), we define differentiation as individualizing student curriculum in online environments.

As such, a distinction can be made between what could be considered vertical or lateral differentiation. Vertical differentiation can be thought of as forms of acceleration when a student is moved to a new unit, or course (the differentiation counterpart to within-classroom ability grouping). Conversely, lateral treatment can be thought of as when a student is moved to an alternative form of educational methods and content, that is designed to facilitate the same cognitive growth (a counterpart to between-classroom ability grouping).

Although each form of differentiation has advantages (see Beecher & Sweeny, 2008; Tomlinson, 2000), a need exists to examine whether the type of differentiation influences the quality of educational outcomes. Largely, researched has focused on when to use differentiation, suggesting that only when specific conditions are met (such as when the student wishes to advance) would differentiation be applied (Matthews, Richotte, & McBee, 2013). Yet, differences in the use of lateral and vertical differentiation in terms of cost of employment and efficiency in correcting curricular mismatches remains to be examined.

Individually, differentiation is one way statistics education can be improved as a field, but as previously noted, schools are not capable of meeting all needs. As such, a joint approach making use of enrichment and differentiation is ideal to improving educational outcomes.

Table 3. Examples of differentiation, categorized by theorized type

Differentiation	Definition	Type
Advanced placement	College level curricula during high school	Vertical
Combined classes	Combining adjacent grade classes	Mixed
Concurrent enrollment	Enrolled in high school and college	Mixed
Continuous progress	Progressive content constantly advances.	Vertical
Course diversification	Providing students with multiple course types.	Lateral
Credit by examination	Testing out of a course.	Mixed
Curriculum compacting	Modifying unit to suit student needs.	Lateral
Curriculum telescoping	Complete work in shorter time.	Vertical
Early admissions	Entering a grade early / bypassing grade req.	Vertical
Grade skipping	Accelerating student to the next grade.	Vertical
Partial acceleration	Accelerating student within content domain.	Lateral
Self-paced	Student selects pace of learning material.	Mixed

3.2.3. Enrichment and Differentiation for Statistics Cognition

Differentiation can improve statistics cognition by creating a better fit between the student and their curriculum, whereas enrichment could enhance cognitions by providing practice of skills in meaningful contexts. In essence, enrichment helps generate authentic assessments for students, whereas differentiation helps shift content. Importantly, for either differentiation or enrichment to enhance cognition, it must be designed to measure and make use of the development of student cognition.

However, these issues transcend the classroom, and can be applied to scholarly work to help bridge gaps in gifted and remedial statistics education. For example, while traditional gifted and special education programs may have selected students based on ability, differentiation and enrichment based on a cognitive curriculum requires students are selected for their learning trajectories, or rather their rate of cognitive growth, as determined by educational assessments.

Aside from such a practical use, applying the NE model of statistics education (Figure 2 in Chapter 2), also may shed specific insights into how and when enrichment and differentiation should be used. For instance, if encoding is needed for statistical literacy, then students who are gifted or remedial in terms of statistical literacy, may have exhibit differences in their posterior cingulate, as well as other areas that have been linked to the cognitive process such as encoding (see Anderson et al., 2016). Thus, differentiation or enrichment activities aligned with enhancing literacy outcomes could be ideal for these students. However, curricular adjustments are only warranted when accompanied with evidence verifying how the altered structure increased efficiency or effectiveness of cognitive growth.

Although there are few predictions that may be made at this stage, part of developing a model of statistics education (Chapter 2) is to understand something about the very cognitions we view as educational outcomes. For instance, based on the NE model, enrichment activities may be suited when students have needs in their reasoning or thinking skills that exceeds what their curriculum can currently offer because enrichment can help students see skills as having high utility and provides an authentic context for demonstrating their abilities. Conversely, differentiation, which is focused more on the content students have access to, may be more appropriate when literacy skills are causing a curricular mismatch.

With a framework established for how to develop a cognitive curriculum and an outline of ways curriculum can be retroactively adjusted, we can return our focus to understanding these features of quality education as they apply to online learning environments.

4. ONLINE LEARNING IMPLICATIONS: COGNITIVE CURRICULUM IN DIGITAL STATISTICS ENVIRONMENTS

So far, much of the discussion has centered on how to develop curriculum for statistics education, as opposed to statistics education for online courses. This approach was taken because, as noted previously, good educational practices often transcend modality (Shale, 1990). That is, regardless of mode of the mode of the educational environment, quality educational programs can be designed (Allen et al., 2004).

One of the problems with comparing online and face-to-face learning environments is that operationalization for these modes of instruction varies drastically (Zhao et al., 2005). To better combat this question, scholars framed comparative analyses on use of technology, and found that students who use technology score, on average, 12 points higher relative to students in courses that do not use technology (Tamin, Bernard, Borokhovski, Abrami, & Schmid, 2011). Although previous concerns, such as whether course grade is framed on cognitive objectives within these studies, remains, converging literature has suggested that technological advances can improve student learning (Means et al., 2009).

To some degree, online learning may be beneficial because of the way in can be used to create an interactive learning environment (de Bruyn, 2004; Delialioglu & Yildirim, 2007). Interactivity and use of graphic media has consistently been found to enhance learning statistics (Valero-Mora & Ledesma, 2011). The use of statistical software (West, Wu, & Heydt, 2004), Excel™ spreadsheets (Cryer, Should, & Marks, 2001), graphing calculators (Dunham & Dick, 1994), and multimedia (Symanzik & Vukasinovic, 2006), have been viewed as viable ways to enhance statistics education (see Chance et al., 2007).

However, more recently, research has examined the use of technology to deliver instruction, with meta-analytic findings supporting web-based over classroom instruction for declarative knowledge, and finding no differences in delivering procedural knowledge (Sitzmann et al., 2006). For example, in a meta-analysis of 45 studies examining the efficacy of computer-assisted instruction (CAI) across statistics courses, researchers found a small (d = 0.30) advantage CAI (Sosa, Berger, Saw, & Mary, 2010). Although it remains unclear whether the advantage had more to do with the technological resources used within the class, this work suggests that the instructional component of online statistics courses may also afford online courses unique advantages in statistics education.

While this Chapter has focused on generalized considerations in developing a statistics curriculum framed on cognitive objectives, we argue that online environments are beneficial to statistics education for both student learning, and instructional delivery because they can uniquely incorporate the micro and macro-level curricular features discussed in a systematic and scientifically rigorous manner.

4.1. Advantages of an Online Statistics Education

From a microscopic perspective, we have argued that statistics curriculum must be formed on cognitive objectives, tasks, and assessments, are designed to promote transfer of cognition. Although such feats can be achieved in a statistics course, regardless of modality, online statistics environments may offer unique benefits when aligning curriculum, that could enhance both learning and teaching statistics.

4.1.1. Benefits for Educators

We have argued that for educators to deliver a cognitive curriculum that they must collect and assess evidence. However, cognitions are often inter-related (Table 1), scaffolding upon one another to achieve any one task (Figure 1). For instance, recall the example of having students test the null hypothesis. While an educator may adopt a seemingly simple five step process to conduct this analysis, each step involves one or more statistics cognitions. Thus, an educator cannot simply look at whether or not the student calculated the correct result, but must examine their thinking and use of statistics cognition every step of the way. Yet, given the amount of null hypothesis tests a student may have to calculate to become proficient, it may be a burden to expect the educator examine each cognition for each step of every problem.

However, educators have often used online platforms to manage their courses more efficently (Cheang, Kurnia, Lim, & Oon, 2003), designing courses that automate portions of grading. This could be of tremendous value for educators delivering a cognitive curriculum in an online statistics course, because they break down each statistics task so that they can gain insight on student cognition. For example, when developing null hypothesis questions for homework assignments and exams, an educator may begin by having students select the appropriate definition for what a null hypothesis is based on an example. Then, students would move on to selecting the appropriate test statistic, and be asked to select the rationale that best coincides with the test statistic selected. This sort of multi-layered questioning targets specific cognitive processes, and can be downloaded by the educator as a progress report for the course or an individual student.

Indeed, much of what we have discussed suggests that online courses may allow educators to collect more data, at a faster rate, and generate the graphs needed to better understand student learning. For example, as previously discussed, students who answer all literacy questions correctly may demonstrate a ceiling effect, whereby they show no cognitive growth (Figure 2). If the educator took care to categorize each problem on an exam as measuring literacy, reasoning, and thinking, the educator could easily download performance data and quickly see whether a ceiling effect exists for the class or an individual. If such an effect is present for an entire class, this may signal to the educator that they may integrate advanced concepts or expedite their pace. Ceiling effects at the individual level could be used to adjust targeted curriculum for students who may show advanced propensity for understanding statistics.

Additionally, technology readily allows educators to adopt formative assessments (see Beatty & Gerace, 2009), that are key to developing cognitive processes. Because we are concerned with how students grow over time, formative assessments are necessary, as they allow educators to see progress. For instance, if an educator is willing to take the time to plan their course, they could easily implement a system of automated online learning modules designed to give students more practice. Such an approach could even be used to target student motivation.

For example, when developing homework assignments, break the assignment into sections based on topics and cognitions. For example, imagine a student answered 7 out of 10 problems designed to assess statistics literacy for measures of central tendency incorrectly. If a student is unhappy with their grade, tell them that they have one week from the date the assignment was returned to complete a central tendency module, which if completed with a passing grade, will allow them to recoup all points lost on their homework assignments. Such formative practices have been shown to enhance student learning (Black, Harrison, Lee, Marshall, Wiliam, 2004), motivation (Cauley & McMillan, 2010) and engagement (Gikandi, Morrow, Davis, 2011), that are all critical to the success of an online course. Although such assessments can be used without technology, platforms such as Blackboard and Moodle can automatically grade student work, allowing the educator to spend less time grading, and focus more time on examining and refining their curriculum (for a review of automated assessment approaches, see Ala-Mutka, 2005).

In addition to enhance the ease of delivering assessments and grading student performance, online courses also allow educators a variety of data sources that can be used to inform their curriculum. As noted previously, one important requirement when developing a cognitive curriculum is gathering proof that curriculum is well aligned. Although one can consider alignment as an issue existing among the student and their curriculum (see 4.1.2), and educator must also seek alignment among their objectives, assessments, and curriculum (Figure 3).

As displayed in Figure 3, when designing a statistics course an educator will create a series of objectives to enhance statistics literacy, reasoning, and or thinking. The proportion of objectives that are related to literacy, reasoning, or thinking, will vary on the course level and students. For instance, an undergraduate course may focus more on literacy objectives, relative to a second tiered graduate course on ANOVA, that may focus more on reasoning. In a sense, the relative portion of objectives determines the focus of the course curriculum, and signals to students what is important. For instance, if 90% of objectives are based on literacy, then students should expect that the course materials, and assessments, will focus on literacy. We argue that curriculum must be aligned so that the relative frequency of objectives matches the course

Figure 3. Cognitive calibration among learning objectives, content covered, and assessments

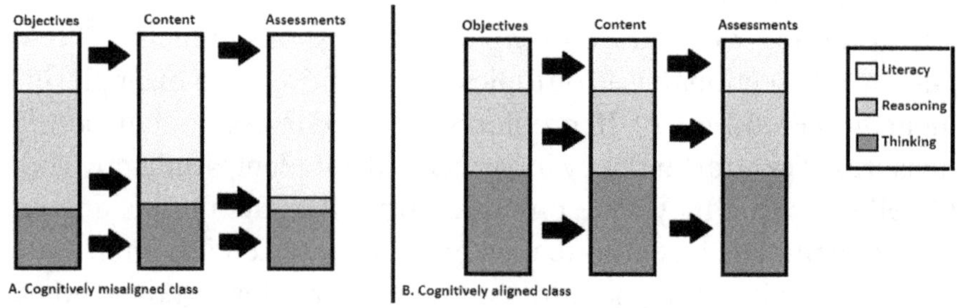

content and types of assessments, and that striving for such alignment can have a positive effect on learning.

For instance, imagine a student who goes to each class, takes notes, and even makes flash cards to help study content. They notice that you spend most of your class defining terms. As such, it would be reasonable for a student to expect that memorizing definitions will allow them to succeed. However, now imagine that only 10% of test items ask for definitions, and the rest ask students to evaluate and synthesize constructs. The curriculum did not afford the student the opportunity to practice the skills they would primarily be tested on, and were in a sense, set up to fail. Yet, if educators develop a master list of what cognitions each assessment - or item - assesses, then an educator can examine what proportion of their items, assessments, and assigned duties that gauge each cognition.

In a sense, a primary benefit of teaching with technology is that it allows educators who planned excellent courses to collect immediate data and make real time decisions about their curriculum. Even information such as how long students spend logged in, how long before exams are readings viewed, and an objective number of total interactions a student had, can be used to assess the ability of curriculum to engage students, and whether or not it results in a significant increase of performance on course outcomes.

4.1.2. Benefits for Learners

In addition to benefits of teaching, online statistics education promises a variety of advantages for the learner. Although some advantages may have a lot to with readily accessible tools designed to improve student efficiency, others have more to do with the potential uniquely afforded by virtue of teaching a course online. Indeed, we maintain that online courses are uniquely beneficial when using a cognitive curriculum.

For example, when developing a cognitive curriculum, learning objectives need to be tightly aligned to assessments and learning activities. As a result, a significant amount of thought and planning must go into selecting appropriate objectives, and crafting reliable and valid educational materials. Similarly, researchers have suggested that course structure (Romiszowski & Cheng, 1992), transparency (Eastmond, 1995), and communication ability (Irani, 1998), are all features that can influence the satisfaction students have in online learning. Students thrive in well organized educational environments, and although educators in face-to-face learning environments are capable of planning to this degree, developing an online course requires the educator to follow through with this initiative.

Online courses may also have an advantage of convenience (Bolliger, 2004). Particular in a visually driven field, such as statistics, graphics are commonly thought to enhance student learning (Dambolena, 1986; Gordon & Gordon, 1989). Any educator who has ever asked students to go home and visit a link shared on Blackboard knows that in many instances, students may not follow through. However, in an online course, educators have the advantage of associated hyperlinks into course content, allowing students to access more materials with greater ease. Instead of expecting a student to go home, turn on a computer, and remember to login to an external educational platform, the relevant content is all presented in a single area for the student to review at their leisure.

Additionally, assessments in online statistics environments may be more likely to be authentic relative to assessments in face-to-face environments. For instance, an easy way to determine whether or not a student knows the definition for a standard deviation is to ask them to define the term. In a face-to-face environment, the educator monitors students to ensure that they comply with the rules of academia. However, such monitoring may not be possible in online learning (see Rowe, 2004). As a result, educators may resort to asking students to generate an example of a term, as opposed to defining it, that may gear online assessments toward higher level cognitive processes.

In addition to assessing higher level cognitions, the structure of these types of assessments may be unique (Rovai, 2000). For example, although it is easy to ask someone if they can define the standard deviation, there may be few times in a student's career where they are asked this question in such a formal and explicit manner. However, by designing assessments that are not easy to cheat in online environments, students are being assessed in a more ecologically valid manner. By designing formative assessments appropriate for measuring cognitive growth in an online environment (Garfield & Chance, 2000; Garfield & Gal, 1999), students may be more authentic (Chance, 1997) in that outcomes are assessed in a manner that more closely matches the way their work will be judged outside of academia.

Further, online courses may allow for differentiation to be more precise, automated, and could even be designed similar to computer adaptive tests, that can adjust in difficulty without the students knowledge (see Olsen, 1986; Van der Linden & Glas, 2000).For example, in a study assessing computer adaptive tests in an online mathematics environment, researchers developed an item response model (see Elo, 1978) that estimated student ability, and adjusted the difficulty of items with each response (Klinkenberg, Straatemeier, & Van der Maas, 2011). In addition to affording a better match between student knowledge and presented curriculum, Klinkenberg and colleagues argued that computer adaptive testing showed greater precision in assessing student ability, and found the assessment to be highly reliable and valid.

However, differentiation need not stop in terms of computer adaptive testing. With enough research, scholars could even develop a precise enough computer adaptive curriculum at within the classroom, or throughout the school, that could be framed on both performance and psychological variables of importance. For example, imagine a student dislikes performing, and has a motivational profile suggestive of mastery orientation (Chapter 6). Although this student may want to learn for the sake of learning, another student may have already mastered the material, and demonstrate a performance approach to the task whereby they want to succeed to show others how much they have learned (see DeShon & Gillespie, 2005; Kaplan & Maehr, 2007; Pintrich, 2000). Because the structure of - and materials for - a course can convey specific information about the nature of course goals, it may one day be plausible and worthwhile to adapt curriculum based on psychological features, such as motivational profiles. Indeed, such structural differentiation would remove negative consequences of traditional tracking and grouping methods, as curriculum could be adjusted without the student's knowledge. At its most extreme, colleges could one day require students to sign up for a specific amount of educational hours, and adapt student curriculum to needs, without any sort of grade level or discipline specific label.

Perhaps one of the most enduring benefits of online courses is that the course physically exists in a digital landscape. As a result, the course inherently guides its community toward particular enrichment resources, that students can readily make use of to bolsters their understanding. For example, an educator could easily make a repository of links to websites like Khan Academy (Thompson, 2011) help students reinforce learning material. As of present, Khan Academy even has a specific statistics module, which offers students video examples of how to define and compute terms, as well as interactive activities such as determining the direction of a correlation. In fact, scholars have even used Khan videos to develop a flipped statistics course, whereby students watch lectures at home so that course time can be used to receive real time assistance as students work through statistics problems (Wilson, 2013). Not only did Wilson maintain that this course improved student performance, but found that it positively influenced student attitudes about the instructor and the course.

Because online courses exist in a digital landscape, they also may serve as an implicit portfolio for student work. E-portfolios have been proven to be a viable tool for assessment in online courses (Mason, Pegler, & Weller, 2004), but the fact that interaction can occur openly in message forms, allows educators the ability to have students reflect on things such as participation, and how their responses may have improved in writing style, content, or use of reasoning. Such self-reflective practices can have a positive influence on student learning (Schunk & Zimmerman, 1998), and some have even gone as far to argue that an educators role in online environments is to develop ways for students to interact and reflect (Maor, 2003).

5. SUMMARY: THE VALUE OF AN ONLINE STATISTICS ENVIRONMENT

We have argued that, in order to develop a course that adheres to practices outlined by statistics education scholars (Ben-zvi & Garfield, 2004; Cobb, 1993; Garfield & Gal, 1997; Franklin et al., 2005), curriculum must be framed on cognitive objectives (Prince, 2004). To achieve a cognitive curriculum, a variety of actions must be taken to carefully plan, administer, and evaluate all materials from micro and macro-perspectives. Although it is possible to achieve this in a face-to-face educational environment, we have argued that online environments are uniquely able to facilitate the development, implementation and evaluation of the curriculum.

We also argue that future scholars need to examine how online statistics education differs from offline counterparts. Perhaps because of barriers perceived in the execution and value of online learning (Harasim, 2000; Lloyd, Byrne, & McCoy, 2012), early work focused on establishing the value of online programming by highlighting how online programs are capable of providing a quality education (Bernard et al., 2004). More recently, work has emerged suggesting that web-based courses may have an advantage (Sitzmann et al., 2006), but concerns in the interpretation of student outcomes persist. For instance, although Sitzmann and colleagues found students in online courses performed better on declarative outcomes (facts and definitions), this material may have less to do with increased performance and more to do with a student's ability to look up facts while online (Eplion & Keefe, 2007). This is not to suggest students cheat. Students can use the internet as a resource before sharing their ideas. Indeed, using technology in this capacity could make assessments more authentic. After all, in the real world, when you do not know the answer to something, or cannot recall a formula, you look it up.

One of the major complications in understanding how online courses influence learning relative to face-to-face instruction is that the quality of learning can depend design features such as organization and structure (see Summers et al., 2005). Yet, as outlined throughout this Chapter, the organization and structure will depend on the nature of the course. Because we have advocated for a curriculum oriented on cognitive objectives, only studies that involves classrooms designed on cognitive curriculum are relevant for comparison.

In some sense, a final grade represents a culmination of the work generated by a student, and conveys final information about performance. However, exactly what performance means in depends largely upon the learning objectives the class was built upon. If a course largely develops and assesses behavioral - as opposed to cognitive - objectives, researchers may not be able to aggregate grades across courses. Indeed, an educator or administrator who uses grades derived from learning objectives framed on different dimensions would akin to a researcher who uses an instrument designed to measure self-confidence to instead measure self-efficacy.

Further, if online learning does indeed have an innate benefit for developing and managing a cognitive curriculum, then online courses could have a different relative proportion of cognitive objectives relative to their counterparts. The problem is not in studying whether or not online classrooms result in higher learning outcomes, but rather in having the ability to equate performance outcomes across classes without knowing details about their structure and purpose. Thus, another key area for future researchers to explore is how alignment influences student outcomes, and whether differences in cognitive outcomes exist among modalities.

Further, future researchers will want to explore the nature of special statistics education. That is, explore the selection, implementation, and evaluation of differentiated curriculum and enrichment activities available to students with special needs. As previously noted, by using a cognitive curriculum these needs are inherently framed on differences in cognitive capacities, suggesting that work focus on how cognitive processes may lead to differences in literacy, reasoning and thinking.

Importantly, while those in need of advanced or remedial statistics education may be underserved, online courses that make use of differentiated curriculum, and curate enrichment resources may be an ideal method to facilitate statistics education across the ability bell curve. Although we have focused on discussing how differentiation could be used to realign tasks to students who may find the typical curriculum to be inadequate, researchers may also want to consider ways that differentiation could be used to serve cognitive and affective outcomes. For instance, if a student adopts a mastery approach in their course work (Chapter 6), they may have different needs relative to a student who is performance approach. As such, it may be conceivable to adapt curriculum to the point where it matches student motivational profiles.

While a variety of studies have explored whether online and face-to-face modes of instruction differ, a need to address this question persists. Although there may be a focus on course grade to measure learning, scholars must develop consistent methodologies to explore these effects if they are going to successfully account for environmental features when making comparisons across universities, disciplines, and even among specific courses.

We believe that online education bears the promise for statistics education, and so as to not risk falling behind, we must proactively examine not only how curriculum shapes course structure, but the very impact it has on learning. By using a cognitive curriculum, we will prepare students to develop lifelong statistical cognition skills, because our end goal is not to instill knowledge but to enhance statistics cognition.

REFERENCES

Ala-Mutka, K. M. (2005). A survey of automated assessment approaches for programming assignments. *Computer Science Education*, *15*(2), 83–102. doi:10.1080/08993400500150747

Aliaga, M., Cobb, G., Cuff, C., Garfield, J. B., Gould, R., Lock, R., ... (2005). *Guidelines for assessment and instruction in statistics education (GAISE): College report.* American Statistical Association. Retrieved from: http://www. amstat. org/education/gaise

Allan, S. D. (1991). Ability grouping research reviews: What do they say about grouping and the gifted. *Educational Leadership*, *48*, 60–65.

Allen, M., Bourhis, J., Burrell, N., & Mabry, E. (2002). Comparing student satisfaction with distance education to traditional classrooms in higher education: A meta-analysis. *American Journal of Distance Education*, *16*(2), 83–97. doi:10.1207/S15389286AJDE1602_3

Allen, M., Mabry, E., Mattrey, M., Bourhis, J., Titsworth, S., & Burrell, N. (2004). Evaluating the effectiveness of distance learning: A comparison using meta-analysis. *Journal of Communication*, *54*(3), 402–420. doi:10.1111/j.1460-2466.2004.tb02636.x

Altman, D. G., & Royston, P. (2006). The cost of dichotomising continuous variables. *British Medical Journal*, *332*(7549), 1080. doi:10.1136/bmj.332.7549.1080 PMID:16675816

Anderson, J. R., Pyke, A. A., & Fincham, J. M. (2016). Hidden stages of cognition revealed in patterns of brain activation. *Psychological Science*, *27*(9), 1215–1226. doi:10.1177/0956797616654912 PMID:27440808

Ansalone, G. (2003). Poverty, tracking, and the social construction of failure: International perspectives on tracking. *Journal of Children & Poverty*, *9*(1), 3–20. doi:10.1080/1079612022000052698

Bakker, A. (2004). *Design research in statistics education: On symbolizing and computer tools* (Unpublished doctoral dissertation). Utrecht University, Netherlands.

Baumeister, R. F., Vohs, K. D., & Tice, D. M. (2007). The strength model of self-control. *Current Directions in Psychological Science*, *16*(6), 351–355. doi:10.1111/j.1467-8721.2007.00534.x

Beatty, I. D., & Gerace, W. J. (2009). Technology-enhanced formative assessment: A research- based pedagogy for teaching science with classroom response technology. *Journal of Science Education and Technology*, *18*(2), 146–162. doi:10.1007/s10956-008-9140-4

Beecher, M., & Sweeny, S. M. (2008). Closing the achievement gap with curriculum enrichment and differentiation: One school's story. *Journal of Advanced Academics*, *19*, 502–530. doi:10.4219/jaa-2008-815

Ben-Zvi, D., & Garfield, J. B. (2004). Introduction to statistical literacy, reasoning, and thinking. In D. Ben-Zvi & J. B. Garfield (Eds.), *The challenge of developing statistical literacy, reasoning and thinking* (pp. 1–17). Dordrecht, Netherlands: Kluwer Academic Publishers. doi:10.1007/1-4020-2278-6

Bernard, R. M., Abrami, P. C., Lou, A. Y., Borokhovski, A. W., Wade, L., Wozney, L., & Huang, B. et al. (2004). How does distance education compare with classroom instruction? A meta-analysis of the empirical literature. *Review of Educational Research*, *74*(3), 379–439. doi:10.3102/00346543074003379

Betts, J. R., & Shkolnik, J. L. (2000). The effects of ability grouping on student achievement and resource allocation in secondary schools. *Economics of Education Review*, *19*(1), 1–15. doi:10.1016/S0272-7757(98)00044-2

Beyth-Marom, R., Fidler, F., & Cumming, G. (2008). Statistical cognition: Towards evidence- based practice in statistics and statistics education. *Statistics Education Research Journal*, *7*, 20–39.

Boaler, J., William, D., & Brown, M. (2000). Students experiences of ability grouping— disaffection, polarisation, and the construction of failure. *British Educational Research Journal*, *26*(5), 631–648. doi:10.1080/713651583

Bolliger, D. U. (2004). Key factors for determining student satisfaction in online courses. *International Journal on E-Learning*, *3*, 61–67.

Bonner, S. E. (1999). Choosing teaching methods based on learning objectives: An integrative framework. *Issues in Accounting Education*, *14*(1), 11–15. doi:10.2308/iace.1999.14.1.11

Bradstreet, T. E. (1996). Teaching introductory statistics courses so that nonstatisticians experience statistical reasoning. *The American Statistician*, *50*, 69–78. doi:10.1080/00031305.1996.10473545

Brown, A. L. (1992). Design experiments: Theoretical and methodological challenges in creating complex interventions in classroom settings. *Journal of the Learning Sciences*, *2*(2), 141–178. doi:10.1207/s15327809jls0202_2

Brown, J. S., Collins, A., & Duguid, P. (1989). Situated cognition and the culture of learning. *Educational Researcher*, *18*(1), 32–42. doi:10.3102/0013189X018001032

Bryce, G. R., Gould, R., Notz, W. I., & Peck, R. L. (2001). Curriculum guidelines for bachelor of science degrees in statistical science. *The American Statistician*, *55*(1), 7–13. doi:10.1198/000313001300339879

Carraher, T. N., Carraher, D. W., & Schliemann, A. D. (1985). Mathematics in the streets and in schools. *The British Journal of Developmental Psychology*, *3*(1), 21–29. doi:10.1111/j.2044-835X.1985.tb00951.x

Cauley, K. M., & McMillan, J. H. (2010). Formative assessment techniques to support student motivation and achievement. *The Clearing House: A Journal of Educational Strategies, Issues and Ideas*, *83*(1), 1–6. doi:10.1080/00098650903267784

Chance, B., Ben-Zvi, D., Garfield, J. B., & Medina, E. (2007). The Role of Technology in Improving Student Learning of Statistics. *Technology Innovations in Statistics Education Journal, 1*, 1-24. Retrieved from: http://escholarship.org/uc/item/8sd2t4rr

Chance, B. L. (1997). Experiences with authentic assessment techniques in an introductory statistics course. *Journal of Statistics Education, 5*, 1–15. Retrieved from http://ww2.amstat.org/publications/jse/v5n3/chance.html

Cheang, B., Kurnia, A., Lim, A., & Oon, W. C. (2003). On automated grading of programming assignments in an academic institution. *Computers & Education, 41*(2), 121–131. doi:10.1016/S0360-1315(03)00030-7

Cheatham, M. L. (2000). A structured curriculum for improved resident education in statistics. *The American Surgeon, 66*, 585–588. PMID:10888136

Cobb, G. W. (1993). Reconsidering statistics education: A national science foundation conference. *Journal of Statistics Education, 1*, 1–28. Retrieved from http://ww2.amstat.org/publications/jse/v1n1/cobb.html

Cobb, G. W. (2007). The introductory statistics course: A Ptolemaic curriculum?. *Technology Innovations in Statistics Education, 1*, 1–15.

Coleman, J. (1961). *The adolescent society*. New York, NY: Free Press of Glencoe.

(1998). Conclusions and future directions for academic interventions. InSchunk, D. H., & Zimmerman, B. J. (Eds.), *Self-regulated learning: From teaching to self-reflective practice* (pp. 225–235). New York, NY: Guilford Press.

Cryer, J. D., Should, G. G., & Marks, T. (2001). Problems with using Microsoft Excel for statistics. *Proceedings of the Annual Meeting of the American Statistical Association*, Joint Statistical Meetings, August 5-9.

Cumming, G. (2013). *Understanding the new statistics: Effect sizes, confidence intervals, and meta-analysis*. New York, NY: Routledge.

Dai, D. Y., & Chen, F. (2014). *Paradigms of gifted education: A guide to theory-based, practice-focused research*. Waco, TX: Prufrock Press.

Dai, D. Y., & Sternberg, R. J. (Eds.). (2004). *Motivation, emotion, and cognition: Integrative perspectives on intellectual functioning and development*. Mahwah, NJ: Lawrence Erlbaum Associates.

Dambolena, I. G. (1986). Using Simulation in Statistics Courses. *Collegiate Microcomputer*, *4*, 339–344.

Dawson, N. V., & Weiss, R. (2012). Dichotomizing continuous variables in statistical analysis a practice to avoid. *Medicine Epidemiology & Biostatistics*, *32*, 225–226. doi:10.1177/0272989X12437605 PMID:22457338

de Bruyn, L. L. (2004). Monitoring online communication: Can the development of convergence and social presence indicate an interactive learning environment?. *Distance Education*, *25*(1), 67–81. doi:10.1080/0158791042000212468

Delialioglu, O., & Yildirim, Z. (2007). Students' perceptions on effective dimensions of interactive learning in a blended learning environment. *Journal of Educational Technology & Society*, *10*, 133–146.

DeShon, R. P., & Gillespie, J. Z. (2005). A motivated action theory account of goal orientation. *The Journal of Applied Psychology*, *90*(6), 1096–1127. doi:10.1037/0021-9010.90.6.1096 PMID:16316268

Dietz-Uhler, B., Fisher, A., & Han, A. (2007). Designing online courses to promote student retention. *Journal of Educational Technology Systems*, *36*(1), 105–112. doi:10.2190/ET.36.1.g

Dunham, P. H., & Dick, T. P. (1994). Research on graphing calculators. *Mathematics Teacher*, *87*, 440–445.

Dunlosky, J., Rawson, K. A., Marsh, E. J., Nathan, M. J., & Willingham, D. T. (2013). Improving students learning with effective learning techniques: Promising directions from cognitive and educational psychology. *Psychological Science in the Public Interest*, *14*(1), 4–58. doi:10.1177/1529100612453266 PMID:26173288

Eastmond, D. V. (1995). *Alone but together: Adult distance study through computer conferencing*. Cresskill, NJ: Hampton Press.

Edelson, D. C. (2001). Learning-for-use: A framework for the design of technology-supported inquiry activities. *Journal of Research in Science Teaching*, *38*(3), 355–385. doi:10.1002/1098-2736(200103)38:3<355::AID-TEA1010>3.0.CO;2-M

Edelson, D. C., Gordin, D. N., & Pea, R. D. (1999). Addressing the challenges of inquiry-based learning through technology and curriculum design. *Journal of the Learning Sciences*, *8*(3-4), 391–450. doi:10.1080/10508406.1999.9672075

Eder, D. (1981). Ability grouping as a self-fulfilling prophecy: A micro-analysis of teacher- student interaction. *Sociology of Education*, *54*(3), 151–162. doi:10.2307/2112327

Elo, A. (1978). *The rating of chessplayers, past and present*. New York, NY: Arco Publishers.

Eplion, D. M., & Keefe, T. J. (2007). Practical tips for preventing cheating on online exams. *Faculty Focus*. Retrieved from: http://www.magnapubs.com/issues/magnapubs_ff/4_4/news/600136-1.html

Epple, D., Newlon, E., & Romano, R. (2002). Ability tracking, school competition, and the distribution of educational benefits. *Journal of Public Economics*, *83*(1), 1–48. doi:10.1016/S0047-2727(00)00175-4

Ericsson, K. A., Krampe, R. T., & Tesch-Römer, C. (1993). The role of deliberate practice in the acquisition of expert performance. *Psychological Review*, *100*(3), 363–406. doi:10.1037/0033-295X.100.3.363

Fejgin, N. (1994). Participation in high school competitive sports: A subversion of school mission or contribution to academic goals?. *Sociology of Sport Journal*, *11*(3), 211–230. doi:10.1123/ssj.11.3.211

Franklin, C., Kader, G., Mewborn, D. S., Moreno, J., Peck, R., Perry, M., & Scheaffer, R. (2005). *Guidelines for assessment and instruction in statistics education (GAISE) report: A pre-K-12 curriculum framework*. Alexandria, VA: American Statistical Association. Retrieved from amstat.org/education/gaise/

Fredricks, J. A., & Eccles, J. S. (2010). Breadth of Extracurricular Participation and Adolescent Adjustment Among African-American and European-American Youth. *Journal of Research on Adolescence*, *20*(2), 307–333. doi:10.1111/j.1532-7795.2009.00627.x PMID:22837637

Friel, S. N., & Corwin, R. B. (1990). The Statistics Standards in K-8 Mathematics. *The Arithmetic Teacher*, *38*, 35–39.

Froelich, A. G., & Stephenson, W. R. (2013). How much do M&Ms weigh?. *Teaching Statistics*, *35*(1), 14–20. doi:10.1111/j.1467-9639.2012.00515.x

Gal, I., & Garfield, J. B. (Eds.). (1997). *The assessment challenge in statistics education* (Vol. 12). Amsterdam, Netherlands: IOS Press.

Gall, M. D. (1981). *Handbook for evaluating and selecting curriculum materials*. Washington, DC: Allyn & Bacon.

Gamoran, A. (1987). The stratification of high school learning opportunities. *Sociology of Education*, *60*(3), 135–155. doi:10.2307/2112271

Gamoran, A. (1992). Synthesis of research: Is ability grouping equitable?. *Educational Leadership*, *50*, 11–17.

Gamoran, A., & Mare, R. D. (1989). Secondary school tracking and educational inequality: Compensation, reinforcement, or neutrality?. *American Journal of Sociology*, *94*(5), 1146–1183. doi:10.1086/229114

Garfield, J. B., & Chance, B. (2000). Assessment in statistics education: Issues and challenges. *Mathematical Thinking and Learning*, *2*(1-2), 99–125. doi:10.1207/S15327833MTL0202_5

Garfield, J. B., & Gal, I. (1999). Assessment and statistics education: Current challenges and directions. *International Statistical Review*, *67*(1), 1–12. doi:10.1111/j.1751-5823.1999.tb00377.x

Gikandi, J. W., Morrow, D., & Davis, N. E. (2011). Online formative assessment in higher education: A review of the literature. *Computers & Education*, *57*(4), 2333–2351. doi:10.1016/j.compedu.2011.06.004

Godden, D. R., & Baddeley, A. D. (1975). Context-dependent memory in two natural environments: On land and underwater. *British Journal of Psychology*, *66*(3), 325–331. doi:10.1111/j.2044-8295.1975.tb01468.x

Gordon, F. S., & Gordon, S. P. (1989). Computer graphics simulations of sampling distributions. *Collegiate Microcomputer*, *7*, 185–189.

Gronlund, N. E. (1985). *Stating objectives for classroom instruction* (3rd ed.). New York, NY: Macmillan Publishing Company.

Hagger, M. S., Wood, C., Stiff, C., & Chatzisarantis, N. L. (2010). Ego depletion and the strength model of self-control: A meta-analysis. *Psychological Bulletin, 136*(4), 495–525. doi:10.1037/a0019486 PMID:20565167

Hallinan, M. T. (1994). Tracking: From theory to practice. *Sociology of Education, 67*(2), 79–84. doi:10.2307/2112697

Hanushek, E. A., & Wößmann, L. (2006). Does early tracking affect educational inequality and performance? Differences-in-differences evidence across countries. *The Economic Journal, 116*, C63–C76. doi:10.1111/j.1468-0297.2006.01076.x

Harasim, L. (2000). Shift happens: Online education as a new paradigm in learning. *The Internet and Higher Education, 3*(1-2), 41–61. doi:10.1016/S1096-7516(00)00032-4

Hessling, R. M., & Schmidt, T. J. (2003). Floor effect. In M., Lewis-Beck, A.E., Bryman, & T. Liao (Eds.), *The Sage encyclopedia of social science research methods* (pp. 390-391). Thousand Oaks, CA: Sage Publications.

Irani, T. (1998). Communication potential, information richness and attitude: A study of computer mediated communication in the ALN classroom. *ALN Magazine, 2*, 1-12.

Kaplan, A., & Maehr, M. L. (2007). The contributions and prospects of goal orientation theory. *Educational Psychology Review, 19*(2), 141–184. doi:10.1007/s10648-006-9012-5

Karpicke, J. D., Butler, A. C., & Roediger, H. L. III. (2009). Metacognitive strategies in student learning: Do students practise retrieval when they study on their own?. *Memory (Hove, England), 17*(4), 471–479. doi:10.1080/09658210802647009 PMID:19358016

Kelly, A. V. (2009). *The curriculum: Theory and practice*. Thousand Oaks, CA: Sage Publications.

Kim, J., & Bastedo, M. N. (2016). Athletics, clubs, or music? The influence of college extracurricular activities on job prestige and satisfaction. *Journal of Education and Work, 30*, 1–21. doi:10.1080/13639080.2016.1165341

Kishner, D., & Whitson, J. A. (1997). *Situated cognition, social, semiotic, and psychological perspectives*. Mahwah, NJ: Lawrence Erlbaum Associates.

Klinkenberg, S., Straatemeier, M., & van der Maas, H. L. (2011). Computer adaptive practice of maths ability using a new item response model for on the fly ability and difficulty estimation. *Computers & Education, 57*(2), 1813–1824. doi:10.1016/j.compedu.2011.02.003

Kulik, C. L., & Kulik, J. A. (1982). Effects of Ability Grouping on Secondary School Students: A Meta-Analysis of Evaluation Findings. *American Educational Research Journal, 19*(3), 415–428. doi:10.3102/00028312019003415

Kulik, J. A., & Kulik, C. L. (1984). Effects of Accelerated Instruction on Students. *Review of Educational Research, 54*(3), 409–425. doi:10.3102/00346543054003409

Lloyd, S. A., Byrne, M. M., & McCoy, T. S. (2012). Faculty-perceived barriers of online education. *Journal of Online Learning and Teaching, 8*. Retrieved from: http://jolt.merlot.org/vol8no1/lloyd_0312.htm

Lohman, D. F. (1993). Teaching and testing to develop fluid abilities. *Educational Researcher, 22*(7), 12–23. doi:10.3102/0013189X022007012

Lou, Y., Abrami, P. C., Spence, J. C., Poulsen, C., Chambers, B., & dApollonia, S. (1996). Within-class grouping: A meta-analysis. *Review of Educational Research, 66*(4), 423–458. doi:10.3102/00346543066004423

Maor, D. (2003). The teachers role in developing interaction and reflection in an online learning community. *Educational Media International, 40*(1-2), 127–138. doi:10.1080/0952398032000092170

Marsh, H. W. (1992). Extracurricular activities: Beneficial extension of the traditional curriculum or subversion of academic goals?. *Journal of Educational Psychology, 84*(4), 553–562. doi:10.1037/0022-0663.84.4.553

Martin, A. J. (2008). Enhancing student motivation and engagement: The effects of a multidimensional intervention. *Contemporary Educational Psychology, 33*(2), 239–269. doi:10.1016/j.cedpsych.2006.11.003

Martin, A. J., Mansour, M., Anderson, M., Gibson, R., Liem, G. A., & Sudmalis, D. (2013). The role of arts participation in students academic and nonacademic outcomes: A longitudinal study of school, home, and community factors. *Journal of Educational Psychology, 105*(3), 709–727. doi:10.1037/a0032795

Mason, R., Pegler, C., & Weller, M. (2004). E-portfolios: An assessment tool for online courses. *British Journal of Educational Technology, 35*(6), 717–727. doi:10.1111/j.1467-8535.2004.00429.x

Matthews, M. S., Ritchotte, J. A., & McBee, M. T. (2013). Effects of schoolwide cluster grouping and within-class ability grouping on elementary school students academic achievement growth. *High Ability Studies, 24*(2), 81–97. doi:10.1080/13598139.2013.846251

Mayer, R. E. (2004). Should there be a three-strikes rule against pure discovery learning?. *The American Psychologist, 59*(1), 14–19. doi:10.1037/0003-066X.59.1.14 PMID:14736316

Means, B., Toyama, Y., Murphy, R., Bakia, M., & Jones, K. (2009). *Evaluation of evidence- based practices in online learning: A meta-analysis and review of online learning studies*. Washington, DC: US Department of Education.

Merrotsy, P. (2008). Acceleration. In N. J. Salkind & K. Rasmussen (Eds.), *Encyclopedia of educational psychology* (pp. 4–8). Thousand Oaks, CA: Sage Publications.

Metz, M. L. (2010). Using GAISE and NCTM standards as frameworks for teaching probability and statistics to pre-service elementary and middle school mathematics teachers. *Journal of Statistics Education, 18*, 1–27.

Muraven, M., & Baumeister, R. F. (2000). Self-regulation and depletion of limited resources: Does self-control resemble a muscle?. *Psychological Bulletin, 126*(2), 247–258. doi:10.1037/0033-2909.126.2.247 PMID:10748642

Oakes, J. (1985). *Keeping track*. New Haven, CT: Yale University Press.

Oakes, J. (1986). Tracking, inequality, and the rhetoric of reform: Why schools don't change. *Journal of Education, 168*, 60–80.

Oakes, J. (1987). Tracking in secondary schools: A contextual perspective. *Educational Psychologist, 22*(2), 129–153. doi:10.1207/s15326985ep2202_3

Olsen, J. B. (1986). *Comparison and equating of paper-administered, computer-administered and computerized adaptive tests of achievement*. Unpublished manuscript. ERIC Document Reproduction Service No, ED274714.

Opdenacker, C., Fierens, H., Brabant, H. V., Sevenants, J., Spruyt, J., & Slootmaekers, P. J. et al.. (1990). Academic performance in solving chemistry problems related to student working memory capacity. *International Journal of Science Education, 12*(2), 177–185. doi:10.1080/0950069900120206

Park, G., Lubinski, D., & Benbow, C. P. (2013). When less is more: Effects of grade skipping on adult STEM productivity among mathematically precocious adolescents. *Journal of Educational Psychology, 105*(1), 176–198. doi:10.1037/a0029481

Pfaff, T. J., & Weinberg, A. (2009). Do hands-on activities increase student understanding? A case study. *Journal of Statistics Education, 17*, 1–34. Retrieved from https://ww2.amstat.org/publications/jse/v17n3/pfaff.pdf PMID:21603584

Pinar, W. F., Reynolds, W. M., Slattery, P., & Taubman, P. M. (1995). *Understanding curriculum: An introduction to the study of historical and contemporary curriculum discourses*. New York, NY: Peter Lang Publishing.

Pintrich, P. R. (2000). Multiple goals, multiple pathways: The role of goal orientation in learning and achievement. *Journal of Educational Psychology, 92*(3), 544–555. doi:10.1037/0022-0663.92.3.544

Quintana, C., Reiser, B. J., Davis, E. A., Krajcik, J., Fretz, E., Duncan, R. G., & Soloway, E. et al. (2004). A scaffolding design framework for software to support science inquiry. *Journal of the Learning Sciences, 13*(3), 337–386. doi:10.1207/s15327809jls1303_4

Reis, S. M., & Renzulli, J. S. (1995). Curriculum compacting: A systematic procedure for modifying the curriculum for above average ability students. *The Journal of the California Association for the Gifted, 26*, 27–32.

Remsburg, A. J., Harris, M. A., & Batzli, J. M. (2014). Statistics across the curriculum using an iterative, interactive approach in an inquiry-based lab sequence. *Journal of College Science Teaching, 44*(02), 72–81. doi:10.2505/4/jcst14_044_02_72

Renzulli, J. S., Smith, L. H., & Reis, S. M. (1982). Curriculum compacting: An essential strategy for working with gifted students. *The Elementary School Journal, 82*(3), 185–194. doi:10.1086/461256

Rogers, K. B. (1991). *The relationship of grouping practices to the education of the gifted and talented learner: Research-based decision making series.* Storrs, CT: The National Research Center on the Gifted and Talented.

Rogers, K. B. (1993). Grouping the gifted and talented: Questions and answers. *Roeper Review*, *16*(1), 8–12. doi:10.1080/02783199309553526

Rogers, K. B. (2002). Grouping the gifted and talented: Questions and answers. *Roeper Review*, *24*(3), 103–107. doi:10.1080/02783190209554140

Rogers, K. B. (2007). Lessons learned about educating the gifted and talented a synthesis of the research on educational practice. *Gifted Child Quarterly*, *51*(4), 382–396. doi:10.1177/0016986207306324

Romiszowski, A., & Chang, E. C. (1992). Hypertext's contribution to computer-mediated communication: in search of an instructional model. In M. Giardina (Ed.), *Interactive multimedia learning environments* (pp. 111–130). Berlin: Springer; doi:10.1007/978-3-642-77705-9_10

Rosenthal, R., & Jacobson, L. (1968). Pygmalion in the classroom. *The Urban Review*, *3*(1), 16–20. doi:10.1007/BF02322211

Rothbauer, P. (2008). Triangulation. In L.M., Given (Ed.), T*he SAGE encyclopedia of qualitative research methods* (pp. 893-894). Thousand Oaks, CA: Sage Publications.

Rouncefield, M. (1994). Recent changes in statistics A-Level: Personal reflections. *Teaching Mathematics and Its Applications*, *13*(3), 97–100. doi:10.1093/teamat/13.3.97

Rovai, A. P. (2000). Online and traditional assessments: What is the difference?. *The Internet and Higher Education*, *3*(3), 141–151. doi:10.1016/S1096-7516(01)00028-8

Rowe, N. C. (2004). Cheating in online student assessment: Beyond plagiarism. *On-Line Journal of Distance Learning Administration, 7*. Retrieved from: http://calhoun.nps.edu/handle/10945/36015

Rubovits, P. C., & Maehr, M. L. (1971). Pygmalion analyzed: Toward an explanation of the Rosenthal-Jacobson findings. *Journal of Personality and Social Psychology*, *19*(2), 197–203. doi:10.1037/h0031526

Saleh, M., Lazonder, A. W., & De Jong, T. (2005). Effects of within-class ability grouping on social interaction, achievement, and motivation. *Instructional Science, 33*(2), 105–119. doi:10.1007/s11251-004-6405-z

Sankar-DeLeeuw, N. (1999). Gifted preschoolers: Parent and teacher views on identification, early admission and programming. *Roeper Review, 21*(3), 174–179. doi:10.1080/02783199909553957

Scheaffer, R. L., & Jacobbe, T. (2014). Statistics education in the K–12 schools of the United States: A brief history. *Journal of Statistics Education, 22*, 1–14. Retrieved from http://ww2.amstat.org/publications/jse/v22n2/scheaffer.pdf

Schunk, D. H., Pintrich, P. R., & Meece, J. R. (2007). *Motivation in education: Theory, research, and applications* (3rd ed.). Upper Saddle River, NJ: Merrill/ Prentice Hall.

Schwartz, T. A. (2014). Flipping the statistics classroom in nursing education. *The Journal of Nursing Education, 53*(4), 199–206. doi:10.3928/01484834-20140325-02 PMID:24654591

Seow, P. S., & Pan, G. (2014). A literature review of the impact of extracurricular activities participation on students academic performance. *Journal of Education for Business, 89*(7), 361–366. doi:10.1080/08832323.2014.912195

Shadish, W. R., Cook, T. D., & Campbell, D. T. (2002). Statistical conclusion validity and internal validity. In *Experimental and quasi-experimental designs for generalized causal inference* (pp. 45–48). New York, NY: Houghton Mifflin Company.

Shale, D. (1990). Toward a reconceptualization of distance education. In M. G. Moore (Ed.), *Contemporary issues in American distance education* (pp. 333–343). Oxford, UK: Pergamon Press.

Shernoff, D. J., & Vandell, D. L. (2007). Engagement in after-school program activities: Quality of experience from the perspective of participants. *Journal of Youth and Adolescence, 36*(7), 891–903. doi:10.1007/s10964-007-9183-5

Shulruf, B. (2010). Do extra-curricular activities in schools improve educational outcomes? A critical review and meta-analysis of the literature. *International Review of Education, 56*(5-6), 591–612. doi:10.1007/s11159-010-9180-x

Sitzmann, T., Kraiger, K., Stewart, D., & Wisher, R. (2006). The comparative effectiveness of web-based and classroom instruction: A meta-analysis. *Personnel Psychology*, *59*(3), 623–664. doi:10.1111/j.1744-6570.2006.00049.x

Slavin, R. E. (1987). Ability grouping and student achievement in elementary schools: A best- evidence synthesis. *Review of Educational Research*, *57*(3), 293–336. doi:10.3102/00346543057003293

Slavin, R. E. (1990). Achievement effects of ability grouping in secondary schools: A best- evidence synthesis. *Review of Educational Research*, *60*(3), 471–499. doi:10.3102/00346543060003471

Slavin, R. E. (1991). Are cooperative learning and untracking harmful to the gifted. *Educational Leadership*, *48*, 68–71.

Sørensen, A. B., & Hallinan, M. T. (1986). Effects of ability grouping on growth in academic achievement. *American Educational Research Journal*, *23*(4), 519–542. doi:10.3102/00028312023004519

Sosa, G. W., Berger, D. E., Saw, A. T., & Mary, J. C. (2011). Effectiveness of computer-assisted instruction in statistics a meta-analysis. *Review of Educational Research*, *81*(1), 97–128. doi:10.3102/0034654310378174

Steenbergen-Hu, S., & Moon, S. M. (2011). The effects of acceleration on high-ability learners: A meta-analysis. *Gifted Child Quarterly*, *55*(1), 39–53. doi:10.1177/0016986210383155

Streiner, D. L. (2002). Breaking up is hard to do: The heartbreak of dichotomizing continuous data. *Canadian Journal of Psychiatry*, *47*(3), 262–266. doi:10.1177/070674370204700307 PMID:11987478

Summers, J. J., Waigandt, A., & Whittaker, T. A. (2005). A comparison of student achievement and satisfaction in an online versus a traditional face-to-face statistics class. *Innovative Higher Education*, *29*(3), 233–250. doi:10.1007/s10755-005-1938-x

Swiatek, M. A., & Benbow, C. P. (1991). Ten-year longitudinal follow-up of ability-matched accelerated and unaccelerated gifted students. *Journal of Educational Psychology*, *83*(4), 528–538. doi:10.1037/0022-0663.83.4.528

Symanzik, J., & Vukasinovic, N. (2006). Teaching an introductory statistics course with CyberStats, an electronic textbook. *Journal of Statistics Education, 14*. Retrieved from: https://ww2.amstat.org/publications/JSE/v14n1/symanzik.html

Tamim, R. M., Bernard, R. M., Borokhovski, E., Abrami, P. C., & Schmid, R. F. (2011). What forty years of research says about the impact of technology on learning a second-order meta-analysis and validation study. *Review of Educational Research, 81*(1), 4–28. doi:10.3102/0034654310393361

Tanner, D., & Tanner, L. N. (1980). *Curriculum development: Theory into practice*. New York, NY: Macmillan.

Thompson, C. (2011). How Khan Academy is changing the rules of education. *Wired Magazine, 126*, 1–5.

Tieso, C. L. (2003). Ability grouping is not just tracking anymore. *Roeper Review, 26*(1), 29–36. doi:10.1080/02783190309554236

Tintle, N., VanderStoep, J., Holmes, V. L., Quisenberry, B., & Swanson, T. (2011). Development and assessment of a preliminary randomization-based introductory statistics curriculum. *Journal of Statistics Education, 19*. Retrieved from: https://ww2.amstat.org/publications/jse/v19n1/tintle.pdf

Tomlinson, C. A. (2000). What is differentiated instruction. In C.M. Callahan & H.L. Herthberg-Davis (Eds.), *Fundamentals of gifted education: Considering multiple perspectives* (pp. 287-300). New York, NY: Routledge.

Tomlinson, C. A. (2003). Deciding to teach them all. *Educational Leadership, 61*, 6–11.

Valero-Mora, P. M., & Ledesma, R. D. (2011). Using interactive graphics to teach multivariate data analysis to psychology students. *Journal of Statistics Education, 19*, 1–19. Retrieved from http://ww2.amstat.org/publications/jse/v19n1/valero-mora.pdf

Van der Linden, W. J., & Glas, C. A. (Eds.). (2000). *Computerized adaptive testing: Theory and practice*. Dordrecht: Kluwer Academic. doi:10.1007/0-306-47531-6

Vygotsky, L. S. (1986). *Thought and language (E. Hanfmann & G. Vakar, Trans.)*. Cambridge, MA: MIT Press.

Weingartner, H., Miller, H., & Murphy, D. L. (1977). Mood-state-dependent retrieval of verbal associations. *Journal of Abnormal Psychology, 86*(3), 276–284. doi:10.1037/0021-843X.86.3.276 PMID:874186

Weinstein, Y., McDermott, K. B., & Roediger, H. L. III. (2010). A comparison of study strategies for passages: Rereading, answering questions, and generating questions. *Journal of Experimental Psychology. Applied, 16*(3), 308–316. doi:10.1037/a0020992 PMID:20853989

West, W., Wu, Y., & Heydt, D. (2004). An introduction to StatCrunch 3.0. *Journal of Statistical Software, 9*(5). doi:10.18637/jss.v009.i05

Wilks, S. S. (1951). Undergraduate statistical education. *Journal of the American Statistical Association, 46*(253), 1–18. doi:10.1080/01621459.1951.10500763

Wilson, S. G. (2013). The flipped class: A method to address the challenges of an undergraduate statistics course. *Teaching of Psychology, 40*(3), 193–199. doi:10.1177/0098628313487461

Wishart, J. (1939). Some aspects of the teaching of statistics. *Journal of the Royal Statistical Society, 102*(4), 532–564. doi:10.2307/2980168

Zhao, Y., Lei, J., Yan, B., Lai, C., & Tan, H. S. (2005). What makes the difference? A practical analysis of research on the effectiveness of distance education. *Teachers College Record, 107*(8), 1836–1884. doi:10.1111/j.1467-9620.2005.00544.x

Chapter 5
Affect in Statistics Cognition

ABSTRACT

This chapter focuses on affective process in statistics education which could influence cognitive development. We begin with a discussion of short-term high-intense affective features (emotions) and use the most well researched construct in computer and statistics education (anxiety) to illustrate how these processes can influence learning. We then discuss long term low intense affective features (moods) and outline how moods can contribute to statistical attitudes. We argue that affective features must be continuously assessed throughout the entire learning process, and discuss theorize how online learning environments can use the principles of differentiation to enhance affect toward statistics education.

...[F]or the context of learning outside the classroom in a mathematics and statistics course: certain emotional components, such as the cognitive, affective or motivational, could play a more important role in the emotional experience of learning... - Niculescu et al., 2015, p. 452

1. INTRODUCTION

Affect, defined as the experience of feeling emotions (Hogg, Abrams, & Martin, 2010), is of central interest in psychological research (Blaney, 1986; Christianson, 2014; Lewis, Haviland-Jones, Barrett, 2010; Russell, 2003),

DOI: 10.4018/978-1-5225-2420-5.ch005

and has emerged as a powerful explanatory variable in statistics education (Niculescu, et al., 2015). Yet despite its importance, research has only just begun to scratch the surface of emotions can influence learning statistics in online environments (DeVaney, 2010; Ramirez, Schau, & Emmioglu, 2012; Tudor, 2006).

Such studies have repeatedly noted differences among affective variables (Niculescu, et al., 2015), with emphasis studying differences in emotions - such as anxiety (Onwuegbuzie & Wilson, 2012) and attitudes (Emmioglu & Capa-Aydin, 2012). As such, this Chapter discusses the role of affective variables in statistics cognition.

While acknowledging the need to understand how affective processes influence cognitive outcomes (Clore & Schnall, 2005; Giner-Sorolla, 1999; Kim, Lim, & Bhargava, 1998), the exact nature of this relationship is still debated (Duncan & Barrett, 2007; Forgas, 2008). Yet for our purposes, what really matters is how affective processes can impact statistics cognitions. To begin addressing this question, we focus on how affect can uniquely influence both generalized (Blaney, 1986) and domain-specific (Mikels, Reuter-Lorenz, Beyer, Fredrickson, 2008; Tremblay, Gardner, Heipel, 2000) cognitive processes, and their related neural sub-systems (Balleine, 2005; Ledoux, 2012; Lane, Nadel, Allen, & Kaszniak, 1999; Van Toller, 1988).

Because affect is such an ambiguous term, we frame out discussion of affect on the most well researched features of affect in statistics education; anxiety and attitudes. Anxiety, which serves as an example of a high intense, but likely short term (situational) emotion (2.1), is a central variable of interest in both online (Chua, Chen, & Wong, 1999) and statistics (Cruise, Cash, & Bolton, 1985) education, with a considerable amount of literature outlining how it influences student learning (2.2). However, anxiety is not simply a construct that can influence learning, but can considered an outcome of statistics education. To further illustrate affective outcomes, we then shift discussion to mood, a relatively long term process with low emotional intensity, to serve as a contrasting example of a long term educational moderator (3.1). Finally, we outline environmental considerations to explore how affect may play a role in self-regulation (4.1) and motivation (4.2) in online environments. We conclude this Chapter with a summary of what we can learn from the study of affective processes in online statistical environments, and where the field may be headed (5).

2. STATISTICAL EMOTIONS

Scholars have found emotion to be an important construct across a wide variety of psychological literatures (Christianson, 2014; Coan & Allen, 2007; Gross, 2013). While cognition generally refers to mental processes such as attention (Taatgen, Van Rijn, & Anderson, 2007) or problem solving (Mayer, 1992), emotions are psychological characterizations of states, such as those elicited by conditioning (LeDoux, 2000; Rolls, 2005), or implicit triggers (Whalen et al., 2004), that are generated from activation of the amygdala (Adolphs, Baron-Cohen, & Tranel, 2002).

Although an entire field of research has devoted itself to understanding emotion (Lewis, et al., 2010), emotions may be of unique importance to learning statistics. For instance, although more and more students are taking statistics courses (Moore, 2001), many students do not learn how to successfully apply their statistics cognitions to daily life (Gal, 2002). To be successful, students must demonstrate a certain degree of efficacy in the choices they make when working with statistical information (Finney & Schraw, 2003). For a student to truly master statistics, they must feel comfortable with how statistics concepts are defined and used.

Feeling comfortable with statistical operations has much to do with the level of anxiety the student feels (Perepiczka, Chandler, & Becerra, 2011; Zeidner, 1991). Indeed, research has noted that students who perceive themselves as having low competence, intellectual ability, and creativity, tend to have higher levels of statistics anxiety (Onwuegbuzie, 2000). Thus, in order to determine how to craft an emotionally responsible online statistics education environment, we will review perhaps the most well researched affective variable in statistics education, statistics anxiety.

2.1. Statistical Anxiety

For decades scholars have noted that statistics courses tend to induce high levels of anxiety among students (Caine, Centa, Doroff, Horowitz, & Wisenbaker, 1978; Schacht & Stewart, 1990). Defined as a worry, intrusive thought, mental disorganization, tension, or arousal when exposed to statistics in any form or at any level (see Onwuegbuzie, DaRos, & Ryan 1997; Zeidner, 1990), statistics anxiety is thought to be relatively pervasive, with as many as 80% of students reporting having felt some anxiety when learning statistics (Onwuegbuzie & Wilson, 2000).

While learning may at times be uncomfortable, scholars have often attempted to study how statistics anxiety forms, and specific methods to reduce its influence (Schacht & Stewart, 1992). For example, Schacht and Stewart (1990) argue that be integrating humor into their lessons, educators may better engage students and reduce anxiety. However, to fully appreciate ways to attenuate the influence of statistics anxiety in online learning, it is necessary to consider the sources of statistics anxiety, and its possible relationship with computer anxiety.

2.1.1. Sources and Antecedents of Statistics Anxiety

Although the possible sources of statistics anxiety are debated (Hunsley, 1987; Onwuegbuzie & Wilson, 2003), scholars have classified specific antecedents such as confidence (Chamberlain, Hillier, Signoretta, 2015) and perfectionism (Onwuegbuzie & Daley, 1999), that are said to be internal sources of anxiety (Onwuegbuzie & Wilson, 2003). However, socially constructed variables, such as gender (Bradley & Wygant, 1998), as well as individual differences, such as ethnicity (Onwuegbuzie, 1999), have also been found to generate anxiety.

While most research examining the impact of statistics anxiety has focused on its negative influence on classroom performance (Lalonde & Gardner 1993; Onwuegbuzie & Seaman 1995; Onwuegbuzie & Wilson, 2003), an emerging body of work has closely examined the potential of the mediating relationship of statistics anxiety with individual differences, and student outcomes. In their review, Onwuegbuzie and Wilson argue that the Anxiety-Expectation Mediation (*AEM*) model - initially used to assess anxiety in learning a new language (see Lalonde & Gardner, 1993) - has been useful in furthering understanding of statistics anxiety.

Asserting that both anxiety and achievement expectations share a reciprocal relationship, the AEM model predicts that a change in anxiety influences a change in expectation, as the two constructs share a reciprocal relationship. To assess the fit of this model, Onwuegbuzie (2003) conducted a path analysis of graduate students ($n = 130$) enrolled in a statistics course. Results suggested a reciprocal relationship, such that expectation was positively related to anxiety ($r = 0.29$) and that statistics anxiety was negatively related to expectation ($r = -0.43$). Further, this model suggested that statistics anxiety was positively related to research anxiety ($r = 0.56$), and that educational factors such as study habits ($r = -0.14$), number of courses taken ($r = -0.22$), and had negative, albeit small, influence on statistics anxiety.

Onwuegbuzie and Wilson (2003) largely defined these results in the context of Wine's (1980) cognitive model, that was used to suggest that the primary difference between students with high and low anxiety is derived from motivation and attention. That is, when students are evaluated in statistics, anxiety interferes with students ability to encode and focus on statistical language and application, as well as their willingness to continue their work. Such findings are aligned with other research that has found that the presence of statistics anxiety diminishes the ability to interpret or analyze statistical information (Onwuegbuzie, 1997).

Further, the links among statistics anxiety and expectation may further understanding of the relationship between statistics anxiety holds with constructs such as perfectionism (Onwuegbuzie & Daley, 1999), instructor immediacy (Williams, 2010), basic math skills (Fonteyne et al., 2015; Lester, 2016), and even gender (Rodarte-Luna & Sherry, 2008). For example, gender research in mathematics has found that by reminding women of a negative stereotype about their gender's performance in mathematics is enough to cause women to perform less well than men (Schmader, Johns, & Forbes, 2008; Spencer, Steele, & Quinn, 1999), a finding that some scholars attribute to anxiety (Johns, Schmader, & Martens, 2005; Schmader & Johns, 2003; Steele, 1997). Indeed, expectations can play a crucial role in student performance (Cadinu, Maass, Frigerio, Impagliazzo, & Latinotti, 2002) and information processing (Seibt & Förster, 2004).

Although this is simply one framework employed to understand the psychological effects of anxiety on statistics performance (Macher, Paechter, & Papousek 2011; Musch & Broder, 1999; Tremblay, Gardner, Heipel, 2000), scholars have meticulously combed the highly interdisciplinary field of statistics anxiety and have developed refined agendas for the future of the discipline (Chew & Dillon, 2014). We urge future scholars to move beyond the valence models of anxiety, and examine anxieties relationship with other emotions from an appraisal perspective.

For example, imagine a student anxious because they do not know how well they are performing in their class. They do not get regular corrective feedback, and their professors has not handed them their last three homework assignments. Now imagine a student who knows exactly where they stand, receives regular feedback, but feels anxious because they doubt their general ability to successful compute statistical values. In this first situation, the affective response conveys uncertainty, and the student may even feel anger toward their professor, as the situation has become motivationally incongruent with their goals (Smith, Haynes, Lazarus, & Pope, 1993). Conversely, in the second situation, there is no specific group or person to blame, and so the flavor the anxiety experiences may be inherently different.

In addition to classifying the features that bring about anxiety from within, it is also necessary to better understand how specific environmental features can generate anxiety. However, in doing so, it is necessary to avoid false dichotomies. Instead of classifying anxiety as a negative feature of statistics education, it may make sense to examine the information conveyed by the features that triggers the anxiety. For instance, traditionally literature has found that students experience anxiety when using computers to learn (Conrad, 2002). Scholars interested in online learning must examine whether online and statistics anxiety are two distinct psychological outcomes, or whether they are derived from similar sources and can be alleviated using the same techniques.

2.1.2. Environmental Triggers: Sources and Antecedents of Computer Anxiety

Like statistics courses, online environments may come with expectations that can also bring about anxiety (Conrad, 2002). Similar to statistics anxiety (Onwuegbuzie & Wilson, 2003), the theoretical basis of computer anxiety can widely vary by scholarly framework (Appelbaum & Primmer, 1990; Howard & Smith, 1986), and while models for computer anxiety may vary in the magnitude of relationships antecedents, there is general consensus that variables such as age, gender, and experience are likely to play central roles its development (Chua et al., 1999).

Such demographic variables are thought to have a meaningful relationships with computer anxiety, and can ultimately influence the attitudes toward the technology used (Igbaria & Parsuraman 1989). Igbaria and Parsuraman's model asserted significant relationships between trait anxiety, and external locus of control, suggesting that a possible interaction among trait-anxious students who are placed in environments that affords little control (for a discussion of state vs. trait constructs, see Spielberger, Gorsuch, Lushene, Vagg, & Jacobs, 1983). However, there was little discussion on the theory that would drive these variables as causal mechanisms.

For instance, although a demographic variable such as education may have a significant, albeit weak, negative relationship with computer anxiety ($r = -0.23$; Igbaria & Parasuraman, 1989), suggesting that as education increases, computer anxiety decreases, the authors did not explore why this might be the case. Perhaps students who had more education were more likely to have exposure to computers, and if experiences are kept largely positive, students with more education may be somewhat more likely to develop an immunity to computer anxiety. Conversely, one could also imagine that people who are trait anxious may be less likely to apply or enroll in a degree program at a college. In this instance, the relationship between education and computer anxiety would be a farce, as the weak relationship would have more to do with a bias self-selected sample in the classroom.

Additionally, the source of computer anxiety could also be seen to vary in scope (Cazan, Maican, & Cocoradă, 2016; Popovich, Gullekson, Morris, & Morse, 2008; Powell, 2013). For example, some scholars argue that the primary source of computer anxiety is unfamiliarity with computers, which creates fear of looking foolish, or breaking the device (Chuo, Tsai, Lan, & Tsai, 2011). However, other scholars have attempted to compartmentalize the derivation of computer anxiety, arguing for upwards of eight distinct sources (Doronina, 1995). While Doronina acknowledges that fear of causing damage to a computer was a potential cause of computer anxiety, they argued that fear of looking foolish was a distinct dimension, as were fears of unfamiliar tasks.

Regardless of these discrepancies, there is a general consensus that trait anxiety, negative experiences with technology, and low self-efficacy are among the central sources of computer anxiety (Cazana, et al., 2016; Chua et al., 1999; Thatcher & Perrewe, 2002), with modern scholarly work generally classifying the sources of computer anxiety as either derived from personal characteristics or environmental interactions (Mullins & Saberwal, 2014). While many sources may be situationally specific, or interact with individual differences, meta-analytic research has suggested that the most robust antecedents of computer anxiety are trait anxiety ($Z = 0.25$), openness to experience ($Z = -0.25$), and neuroticism ($Z = 0.24$; Maricutoiu, 2014).

Although the consequences for computer anxiety are vast (Chua et al., 1999), several outcomes appear to be robust throughout the literature (Maricutoiu, 2014). For example, meta-analytic findings have suggested that computer anxiety leads to lower ratings of ease of using a computer ($Z = -0.39$) as well as a lower intention of using computers in the future ($Z = -0.26$). These outcomes are of great importance, as scholarly work has demonstrated that perceptions of ease of use of computers can shape effectiveness of e-learning (Jashapara & Tai, 2006).

Perceptions aside, research has also noted that individuals with high degrees of computer anxiety perform less well than their counterparts in their coursework (Mahar, Henderson, & Deane, 1997; Marcoulides, 1988), with some scholars even asserting that computer anxiety can completely disrupt successful e-learning (Fuller, Vician, & Brown, 2006). Further, others have argued that computer anxiety can lead to negative beliefs about computers (Beckers & Schmidt, 2001), reduce student motivation (Saadé, Kira, & Nebebe, 2013), and even reduce an educators propensity of using computer supported education (Celik & Yesilyurt, 2013).

However, scholars examining computer anxiety have largely focused on internal enduring trait like focuses, relative to the more situationally explored statistical anxiety research (Onwuegbuzie & Wilson, 2003). Some individual differences, such as those with higher trait anxiety demonstrating higher computer anxiety (Igbaria & Parasuraman, 1989) are intuitive, albeit difficult to detect and combat. Whereas other individual differences, such as gender (Broos, 2005), ethnicity (McInerney, McInerney, & Sinclair, 1994), socioeconomic status (Bozionelos, 2004), and age (Cambre & Cook, 1987), have all been shown to interact with student outcomes. While we maintain that there are important distinctions among the understanding and use of these types of individual differences (Chapter 6), there is enough literature to consider how these indicators can be used to explain variance in online environments.

Regardless of the model of anxiety endorse, scholars agree, that computer anxiety can have harmful consequences on student learning (Buche, Davis, & Vician, 2007; Fuller et al., 2006; Martocchio, 1994) and teacher preparation (McInerney et al., 1994; Russell & Bradley, 1997; Woodrow, 1992). As an educational psychological construct, challenges exist in parsing out the causal mechanism(s) behind computer anxiety, and how they may individually - or jointly - influence student outcomes.

2.2. A Comparison Statistics and Computer Anxieties

Although thc thcorctical basis for both statistics and computer anxiety can vary widely, pinning down their exact source is a difficult task. In part, this may be due to their distinct historical roots and relative focal points of where anxiety is said to be derived from.

From a historical roots perspective, both statistics anxiety (Feinberg & Halperin, 1978), and online education (for review see, Harting & Erthal, 2005) have been researched for well over thirty years. However, many of the topics discussed in online education mirror those relevant to early discussions of the merits of distant educational programs, such as correspondence schools

(Rogers, 1902; Tawney, 1943). However, a unified field of statistics education has only recently emerged (Ben-Zvi & Garfield, 2008), leading scholars in statistics education to take a slightly different approach to examining anxiety, relative to scholars interested in computer anxiety.

Perhaps because of these distinct roots, both lines of research have had a slightly different focus in understanding each form of anxiety. For instance, statistics anxiety is said to be derived from motivational and attentional deficits (Onwuegbuzie & Wilson, 2003), with a reciprocal relationship between anxiety and expectations (Lalonde & Gardner, 1993; Onwuegbuzie & Wilson, 2003). As such, statistics anxiety is often examined as a temporary state, likely arising when a student is placed in a high-stakes educational environment, such as when taking an exam (Onwuegbuzie & Wilson, 2003). Conversely, computer anxiety has been less frequently viewed through the context of attentional and motivational constructs, with scholars largely framing its study in terms of negative experiences with technology and low self-efficacy (Hackbarth, Grover, & Mun, 2003; Wilfong, 2006), with considerable work focusing on the relationship between anxiety as a trait characteristics with situational triggers (Thatcher & Perrewe, 2002).

Although such a distinction may be the product of the way the studies in each field have been framed, it seems likely that students in online statistics courses may experience a rise in anxiety on two fronts. Namely, the discipline may create a rise in state-level anxiety, with the technology negatively interacting with trait-level anxiety. In an online statistics environment, computer anxiety is situated in the environment, suggesting that the *AEM* model (Lalonde & Gardner, 1993) may be of help in understanding the mechanism behind computer anxiety. Because the *AEM* model predicts a reciprocal relationship between anxiety and achievement expectations, self-efficacy can be said to play a central role in combating statistics anxiety. Similarly, both anxieties are described as coming from negative experiences with, especially among those with low self-efficacy (Beckers & Schmidt, 2003; Brosnan, 1998; Finney & Schraw, 2003; Hsu, Wang, & Chiu, 2009; Igbaria & Iivari, 1995; Onwuegbuzie, 1997; Zeidner, 1991).

Although the trait-state distinction among computer and statistics anxiety is helpful in examining how students in online statistics environments could face anxiety on state and trait levels, a need exists to better understand computer anxiety as a state and statistics anxiety as a trait level feature. After all, one reason why such a distinction exists is because the tools used to assess statistics anxiety tend to be more situational (state), the test anxiety sub-dimension of statistics anxiety being the highest loading (see Fabrigar,

Wegener, MacCallum, & Strahan, 1999; Tucker & Lewis, 1973) on factor analyses (Onwuegbuzie, 1998) of the statistics anxiety rating scale (Cruise et al., 1985). Because instruments measuring statistics anxiety tend to focus on state components, researchers are more poised to understand the situational features of anxiety.

As noted in our discussion about the source of affect, it may not necessarily matter the degree statistics and computer anxiety are state or trait, as such a classification is a false dichotomy. Trait and state features are inherently linked, and future research in online statistics education needs to carefully parse apart the purported detriments on student outcomes in terms of the valence and appraisals of triggers.

For example, it may help to consider anxiety levels like thresholds that can vary by situations. Although some may be more predisposed to be anxious (Edwards & Trimble, 1992), it is only under the right circumstances that anxiety may pose a threat (Putwain, Woods, & Symes, 2010). People may have their anxiety thresholds set at different levels (individual personality differences), but it still may take situational features to act as a trigger for anxiety (Baloğlu, 1999). Because computer and statistics anxiety can coincide in an online statistics environment, there may be particular concern regarding the dual impact of these anxieties.

As discussed, statistics and online courses both share some unfavorable perceptions among students, and these perceptions may even be of significance difference when it comes to motivation, self-regulation, and overall course satisfaction (Pekrun, Goetz, Titz, & Perry, 2002; Summers, Waigandt, & Whittaker, 2005). Because these anxieties may have a dual negative influence on educational outcomes, it is important to consider how anxiety can be attenuated, regardless of where it may reside.

2.2.1. Reducing Statistics and Computer Anxieties with Efficacy

Because there is some concern that anxiety could doubly impact educational outcomes in online environments due to the negative effect and prevalence of both computer and statistics anxieties, it is necessary to consider ways both forms of anxiety can be jointly curtailed. For example, statistics anxiety has a moderate negative relationship with self-efficacy ($r = -0.39$; Baloğlu, 2002). Similar trends have identified with computer anxiety (Beckers & Schmidt,

2001), which has been found to have a moderate negative relationship with physical arousal ($r = -0.37$), and small, but positive, relationships with both affective feelings ($r = 0.29$) and positive beliefs ($r = 0.22$). Importantly, fear seems to be a major component within both forms of anxiety, regardless as to whether a student is worried about looking foolish when using statistical software (Chuo et al., 2011) or their use of statistical language (Onwuegbuzie, 1997). If anxiety is associated with fear, it stands to reason that increasing self-efficacy (see Bandura, 1997), is a viable solution naturally lending itself to combating anxiety.

Although the ways self-efficacy can reduce anxiety could vary (Badura, 1994; Pajares, 1996), it is thought to influence behavior via cognitive, motivational, and selection processes (Bandura, 1997). Of particular interest to the discussion at hand is how self-efficacy is can influence the cognitive processes disrupted by anxiety (Bandura, 1997), with a slew of studies finding reciprocal relationships between self-efficacy and anxiety on student performance (Bandura, 1982; 1988; Brosnan, 1998; Strecher, DeVellis, Becker, & Rosenstock, 1986). Arguably, self-efficacy may be able to allow students to better handle environmental triggers toward trait anxiety, while helping them avoid state anxiety. Indeed, meta-analyses have repeatedly found support for the positive relationship of self-efficacy on student performance (Brown, Lent, Larkin, 1989; Shell, Colvin, & Bruning, 1995), with general support for the belief that anxiety is mediated by self-efficacy (see Pajares, 2003).

Evidence presented throughout this Chapter is largely in line with the hypothesis that self-efficacy is mediated by anxiety. For example, as previously discussed, measures of statistics anxiety (Baloğlu, 2002) and computer anxiety (Beckers & Schmidt, 2001) are negatively related to self-efficacy. However, despite their apparent link, little research exists on using self-efficacy interventions (Gist & Mitchell, 1992; Pajares, 1996; Schunk & Ertmer, 2000) in online statistical environments. Although direct examinations of self-efficacy as a means to combat both computer and statistics anxieties have yet to be seen by these authors, meta-analytic findings have suggested moderate positive relationships between self-efficacy, student performance ($r = 0.38$) and persistence ($r = 0.34$) for students regardless of their ability or grade level (Multon, Brown, & Lent, 1991). Indeed, some research has even found self-efficacy has been found to be the most important predictor of success in online statistics courses (Abd-El-Fattah, 2005).

2.2.2. Adaptively Anxious: The Zone of Proximal Anxiety

Even if enhancing self-efficacy is a targeted solution that could prevent statistics and computer anxiety, it is important to note that components of anxiety may be adaptive. For example, anxious and stressful states can often provide useful information about efficacy beliefs (Pajares, 2005), perhaps shedding light on why students with mid-range levels of statistics anxiety performed better than peers who had higher or lower levels (Keeley, Zayac, & Correia, 2008).

The finding of a curvilinear pattern of statistics anxiety on grades (Keeley et al., 2008), may suggest that it is only students with extremely high or low anxiety thresholds who are primarily at risk. Thus, the goal is not to remove anxiety from the educational environment, but to ensure it works to the student's benefit. Such advice parallels findings, such as those presented by models of pedagogical strategies that typically suggest that educators can reduce anxiety and increase confidence by providing (Chen, 2004); (1) clear and adequate feedback, (2) incentives for good performance, (3) ensure the targeted zone of proximal development (see Vygotsky, 1986) is within the student's ability.

This curvilinear pattern suggests that anxiety can be adaptive. From an evolutionary perspective, anxiety may serve an adaptive function (Marks & Nesse, 1994; Nesse, 1990; 1998). For instance, anxiety can provide useful information about one's efficacy (Pajares, 2005), helping to calibrate future performance. Although scholars have long theorized about developing errorless learning techniques (Skinner, 1953; Terrace, 1963), making mistakes is simply part of learning when the opportunity and information to help students learn from their mistakes is present. When a student is not performing up to their expectations, they may start to experience negative affect. Thus, the educator's role may not be so much in reducing negative affect, than it is providing students with optimally anxious situations and guidance modulating learning strategies.

As such, one could consider the optimal state of anxiety to be within a *zone of proximal anxiety* (ZPA). Similar to the conceptualization of the zone of proximal development, the ZPA exists as a theoretical range of anxiety a student is able to manage in an educationally adaptive manner. For instance, consider Figure 1, that depicts two students with differing state and trait level anxieties. Importantly, student A has a higher degree of trait anxiety, with lower state thresholds. This student may find it particularly challenging to learn

Figure 1. Zone of proximal anxiety; depicted in two students

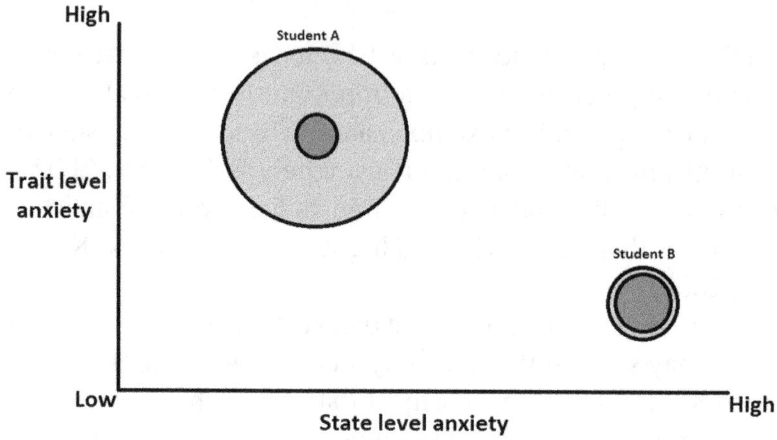

in an online statistics environment, even if their peers felt that the situation was relatively unstressful. Conversely, student B has low trait anxiety, and a high threshold for state level anxiety. This student tends not to get anxious, and even if they may find themselves in a stressful environment.

Yet, it is not necessarily important where students are on this spectrum. Indeed, it may make more sense to think of the degree of adaptive anxiety each student possesses. For instance, as depicted in Figure 1, student A and B also have differential ratios of what they can and cannot do with or without assistance. As with the zone of proximal development, the ZPA posits that some students may have a wider spectrum of adaptive anxiety than others. While student A has a higher level of trait, and lower threshold for state anxiety, they have a wider range of adaptive anxiety, relative to student B, who may be easily triggered by environmental cues.

While too much or too little anxiety can be detrimental to performance (Kelly, 2008), the neurological basis of anxiety suggests that scholars may want to explore how anxiety related regions such as the amygdala (Balleine, 2005; Davis, 1992; Ziabreva, Poeggel, Schnabel, & Braun, 2003), interact with posterior cingulate and polar frontal, ventral visual stream and parietal, prefrontal, and caudate areas (Anderson & Fincham, 2014; Chapter 2) to better understand the dynamic relationship among these constructs, and their influence on learning gains when using a cognitive curriculum (Chapter 4). However, because anxiety is likely sourced from the environment and its biological basis, any approach aimed at adjusting anxiety must take into account of the joint role of the environment and genetics (Chapter 6).

As an emotional state of turmoil, anxiety proves to be of central interest to both researchers and practitioners, looking to improve the quality of their students learning. Indeed,

although anxiety can prompt a motivating state of arousal (Keeley et al., 2008), it also has negative relationships with features we know to be of importance to learning, such as self-efficacy (Chua et al., 1999) and success in online learning environments (Abd-El-Fattah, 2005). Future studies are needed to understand how anxiety manifests in online statistics environments, and how educators can alter design features of online courses to adjust arousal to optimal levels.

3. STATISTICAL MOODS AND THE FORMATION OF STATISTICS ATTITUDES

In addition to the short term high intensity influence of emotion, enduring low intensity affective features play a role in learning statistics. As with anxiety, attitudes toward computers (Kernan & Howard, 1990; Liaw, Huang, & Chen, 2007; Noyes & Garland, 2006; Whitley, 1997) and statistics (Emmioglu & Capa-Aydin, 2012; Evans, 2007; Suanpang, Petocz, & Kalceff, 2004) share different theoretical frameworks that attempt to identify and explain the effect of their antecedents, with both literatures concluding that alternative affective processes, such as mood, can influence one's ability to learn (Bower, 1981; Isen, 1987; Forgas, 1995).

Some of the earliest, and strongest, evidence demonstrating how mood can influence cognition is from literature examining the mood-dependent learning hypothesis (Bartlett & Santrock, 1979; Bower, 1981; Eich & Metcalfe, 1989; Weingartner, Miller, & Murphy, 1977). This hypothesis suggests that one's ability to remember information is enhanced if their mood while encoding is similar to their mood whilc rctricving information.

For instance, in a recent study (Nutt & Lam, 2011), participants diagnosed with bipolar disorder were shown a series of inkblots, and asked to participate in a visual recognition task. Participants performed significantly better when their mood during their testing phase was the same as when they were shown the inkblots.

Importantly, if affect can influence our memory, it may influence other cognitive processes, such as regulation of information processing (Bless, Fiedler, Forgas, 2006; Wyer, Clore, & Isbell, 1999), persuasiveness (Schwarz,

Bless, & Bohner, 1991), and even the types of judgments we make (Forgas & George, 2001; Peters, Västfjäll, Gärling, Slovic, 2006). In a statistics course that is centered on memorizing definitions and formula, mood may not be a concern, but when curriculum is designed to enhance statistics cognitions (Chapter 4), negative moods can disrupt cognitive processes, leading to reduced performance (Haines, Norris, & Kashy, 1996; Heiligenstein, Guenther, Hsu, & Herman, 1996; Pekrun, 1992). Indeed, if cognitive objectives define success by literacy (recall), reasoning (understanding) and thinking (application) outcomes, then students need to process information objectively, and make sound judgments.

3.1. Statistical Moods: How Statistics Attitudes Form

Despite the need to understand statistical mood in online environments, little research has directly measured this relationship. While some research has examined how using humor can lighten the mood of a course (Neumann, Hood, Neumann, 2009), most research focuses on describing (Ramirez et al., 2012) and monitoring attitudes and beliefs (Gal, Ginsburg, & Schau, 1997). However, these works generally approach attitudes as pre-existing states, noting that many students enter the course with experiences computing statistics, mathematics, and gut intuitions about what statistics ought to be (Gal et al., 1997). However, there is a need to empirically examine mood from an educators perspective. Namely, how can educators best create affective learning objectives, and how can they manipulate course content to ensure statistical mood improves over the duration of the term.

Among other suggestions, some practical ways to improve mood are to create an environment where students feel safe to conjecture, are comfortable with confusion, believe their ability to succeed, and are motivated to persist (Gal et al., 1997). However, a need exists to examine how these features can best be achieved. For instance, one way to make students feel comfortable with confusion could be to create a shared questions board students can anonymously submit questions about material, without fear of how they may look to their peers. Yet, is such a feature significantly enhancing that feeling of safety? Perhaps having the instructor post a series of questions to demonstrate that even they are a student of statistics could encourage students to feel comfortable with confusion. Whether such tactics have a meaningful additive effect is an empirical question, and scholars should look toward addressing matters of practice and design efficiency for educators.

We argue that a detailed study of statistical mood is necessary to understand statistics attitudes, as there is evidence cognitive processes are influenced by affect as a means to bias and expedite responses (Vogel & Wanke, 2016). For instance, in their classic study, Schwarz and Clore (1983) documented the *affect heuristic*, a psychological bias where primed emotions attitudes. In their study, participants were called on the phone and asked questions about their life satisfaction. Depending on condition participants attention was drawn to the weather, that was predicted to prime participants responses about their life satisfaction. When participants were reminded of the weather and it was raining, they reported significantly lower life satisfaction then all other conditions.

Aside from making a point about how easily attitudes can be influenced, Schwarz and Clore's (1983) study demonstrated that the affect heuristic can influence attitudes. Although attitudes were not found to differ among online and face-to-face learning environments (Bernard et al., 2004), online college level statistics courses may often require use of additional technology (whether it is in the form of applets, or calculation programs like Excel™), above and beyond what other online courses require. While learning any additional technology should exist as part of the course learning objectives (Chapter 4), repeated exposure to new technology could reinforce negative perceptions of a course, should a student be uncomfortable with new technology. Indeed, comfort with technology has been shown to be an important indicator of student attitudes toward online learning (Drennan, Kennedy, & Pisarski, 2005).

Courses designed to enhance statistics cognitions must adopt methods to enhance statistics motivation. Before this can be achieved, scholars must empirically establish antecedents of statistical moods, and determine whether or not they are predictive of success in online courses designed on a cognitive curriculum (Chapter 4). Further, it is necessary for educators to adopt a course objective of increasing statistical mood. To do this, educators must develop methods to continuously assess mood, and demonstrate growth. To facilitate this, theoreticians can look toward developing models of how design features can improve student mood, in such a way that attitudes about statistics are enhanced.

3.2. Prevalent Attitudes About Online Statistics Education

One way to determine how moods can be improved is by examining the types of attitudes that have been shown to exist about statistics in online environments. Luckily, much of the research focused on online statistics education has examined statistics attitudes. One possible reason that the study of affect in online environments may have been an initial focus of online statistics education research is because students tend to perform equally well in online environments (Allen et al., 2004; Cavanaugh, Gillan, Komrey, Hess, & Blomeyer, 2004), prompting scholars to consider other variables that could explain learning in a cognitive oriented statistics course. Indeed, some have even gone so far as to suggest that the only robustly differences observed among online and face-to-face courses tend to be outcomes that are affectual in nature (Summers et al., 2005).

Attitudes are likely a central concern for statistics educators, as studies have repeatedly shown that positive attitudes are linked to student performance (Carlson & Winquist, 2011; Cashin & Elmore, 2005; Tempelaar & Nijhuis, 2007). However, students often come to class with previously formed attitudes about courses, regardless if they had even previously taken a statistics course (Gal et al., 1997), and often, in quantitative courses, student attitudes are less than positive. For example, one study found that students parsed courses into "easier" qualitative language subjects and "harder" quantitative mathematics and statistics subjects (Murtonen, 2005; Murtonen & Lehtinen, 2003; Onwuegbuzie, Leech, Murtonen, & Tähtinen, 2010) Further, as previously noted (2.1), anxiety about statistics is fairly widespread, having been reported by upwards of 80% of all students (Onwuegbuzie & Wilson, 2003).

Similarly, scholars studying attitudes toward online learning have found attitude to be a strong predictor of success. This is especially true for online statistics courses, that make use of statistical software. For example, scholars generated a model that found that computer attitude and statistical software self-efficacy significantly improve the perceived usefulness of the statistical software, and that perceived usefulness in turn positively influenced the learners tendency to use the software (Hsu et al., 2009). Because computer attitudes can influence propensity to use statistical software, it suggests that computer attitudes may influence statistics cognition, above and beyond the influence of attitudes about statistics.

Previous work has argued that attitudes can be improved in both online (Knowles & Kerkman, 2007) and statistics (Dempster & McCorry, 2009) learning environments, but there is a need to assess to what degree these attitudes overlap, and whether they can be improved using the same methods. Such an multi-pronged intervention may rely on utility, as usefulness has been shown to be key to improving attitudes toward statistics in both face-to-face (Evans, 2007) and online (Hsu et al., 2009) settings.

In addition to enhancing performance, improving attitudes about statistics may also help student motivation. While online courses may pose greater challenges to individual student motivation or self-regulation (Williams & Hellman, 2004), improving student attitudes may go a long way in keeping students motivated (Gardner & Lambert, 1972; Liaw et al., 2007; Vankatesh, 1999). Although scholarly research has focused on the psychometric assessment (Dauphinee, Schau, & Stevens, 1997; Kernan & Howard, 1990) and uses of attitudes in education (Perney & Ravid, 1990; Tempelaar, Schim van der Loeff, & Gijselaers, 2007), future work can begin to react to the negative attitudes many students may have about online statistics courses by demonstrating to the student how useful learning statistics can be.

3.3. Understanding Educationally Important Preferences

If given a choice, what type of statistics course might you pick? Presumably, you would select an environment you are *educationally adapted* to. That is, one that uses similar modes of instruction, and techniques for learning, that you are accustomed to, or prefer. However, is there evidence to suggest that abiding by these preferences for modes of instruction and educational techniques results in additive advantages?

Some of the earliest research to examine this question was concerned with how instruction could be modulated to suit the individual needs reflected within student learning (Kolb, 1976; 1999; Leonard & Harris, 1979). Specifically, scholars proposed that students could be classified into different types of learning, each providing insights into the ways students learn best. While models varied in what constituted a learning style, one reductionist example could be as simple as labeling one student as an auditory learner, who prefers to listen to material, versus a visual learner who prefers to see the material they are learner. Over the years, the concept of learning styles has been tested in a variety of theoretical frameworks (Grasha & Riechmann, 1974; Keefe, 1985), with scholars arguing that learning can be improved by matching students with their learning environments that suited their unique styles of learning (Dunn, 2000; Dunn & Dunn, 1978, 1993).

Such a practice makes intuitive sense. After all, most American schools arrange classes by age, presumably because age is an approximate indicator for the types of content and skills that students possess at a given time. Thus, if all of the students within the same class are the same age, it is assumed that they will roughly be ready for to learn the same material, and if not, remedial and gifted educations are available to readjust (Chapter 4). Similarly, the logic of learning styles suggests that within an age category, there are different ways students can take in information, and that arranging based on these profiles will help make education a more meaningful and efficient experience (Dunn & Dunn, 1992).

Despite its widespread application, recent evidence has been critical of learning style theory (Willingham, Hughes, Dobolyi, 2015). For example, in a recent review, researchers defined several theoretical problems with the lack of evidence supporting learning style theory (Pashler, McDaniel, Rohrer, & Bjork, 2008). Arguing that because prior experiments have not indicated a cross-over interaction (Cronbach & Snow, 1977; Snow, 1989) between methods and learner styles for academic performance, that style of learning may not matter.

3.4. Learning Styles in Online Statistics Environments

However, despite this lack of evidence (Pashler et al., 2008), the use of learning styles is still highly prevalent throughout online (Blum, 2005; Diaz & Cartnal, 1999; Terrell & Dringus, 2000), and statistics education research (Garfield 1993; Hubbard, 1997; Magel, 1998). This is not to imply that researchers in statistics and online education should not explore the nuances of learning styles, but rather, to indicate that despite the lack of performance based evidence supporting the use of learning styles, there may be other means to adjusting course features - in terms of curriculum, or anxiety - to match student preferences.

As discussed one of the primary benefits of offering online instruction is the ability to differentiate class structure, and while learning styles may not inherently impact performance (Pashler et al., 2008), there is more to statistics education than performance, especially when it is based in a cognitive curriculum (Chapter 4). Throughout the Chapter, the importance of effectual variables, such as emotion, mood, and attitudes have been stressed, and if students do hold such a strong preference for their modes of learning, offering materials suited to differential styles may not only improve attitudes, but help adjust anxiety to optimal levels. Clearly, research is needed to understand how learning styles could shift affectual responses in online statistics education.

4. ONLINE LEARNING IMPLICATIONS: MOTIVATED COGNITION IN ONLINE STATISTICS ENVIRONMENTS

Affect is one of the few areas of online statistics education to have both separate discipline specific literatures (Chua et al., 199; Onwuegbuzie & Wilson, 2003), as well as joint literature examining the influence of affect in online statistics environments (Niculescu, et al., 2015). Largely, the literature reviewed suggests students in online learning environments can perform as equally well as peers in face-to-face environments (Allen et al., 2004; Cavanaugh et al., 2004). However, affective features can indirectly influence outcomes related to student performance in online statistics environments (Hsu et al., 2009).

To explore how affect influences performance, a variety of scholars have examined student motivation. For example, positive affect can improve self-regulation (Tice, Baumeister, Shmueli, Muraven, 2007), that is often thought to be imperative to learning (Pintrich, 200). If not properly motivated, students may disengage from curriculum and stop learning. Indeed, online courses may have low retention rates (Carr, 2000; O'Brien & Renner, 2002), with some estimating that retention is between 10-20% lower for online courses (Frankola, 2001). Thus, a need exists to understand how students can be motivated in online environments.

Motivation is a diverse interdisciplinary literature in of itself (Schunk, Pintrich, & Meece, 2007), with varied scholarly perspectives. For instance, some scholars have approached motivation from a design perspective (Keller, 1987), arguing that by designing curriculum that engages students attention, is interpreted as relevance, promotes confidence, and keeping students satisfied, students can be motivated to succeed in online learning environments (Huett, Kalinowski, Moller, & Huett, 2008). Conversely, others have examined how motivation can be situated internally, and can prompt students to adopt different goals that may or may not be conducive to their success in digital environments (Lynch & Dembo, 2004). We take the view that motivation is a product of cognition and affect (Dai & Sternberg, 2004), and argue that motivation is the natural result of appropriately adopting and aligning cognitive and affective learning objectives.

However, little research has examined motivation as an outcome in online statistics environments (Summers et al., 2005). For instance, scholars assessing student motivation in online learning have argued that major indicators of motivation and affect influence student learning outcomes (Chen & Jang,

2010; Lim, 2004). As such, scholars have even developed online statistics courses designed to promote mastery as opposed to performance outcomes (see Kreiner, 2006). However, because such courses take a considerable amount of time and effort, and may not result in clear learning advantages, scholars must refine theories of motivation based individual differences (Chapter 6).

Educators may want to routinely measure motivation, along with affective and cognitive learning objectives, as motivation is a continuous outcome, updating throughout the duration of the learning experience, suggesting that motivations may frame executive attentional resources influencing the quality of learning in real time (Baddeley, 1992; Engle & Kane, 2003). A reciprocal relationship among affect, cognition, and motivation is likely to exist, as affective features recursively update attitudes, that as reviewed throughout this Chapter, can influence a variety of educational outcomes.

In addition to exploring how motivation can influence the quality of learning statistics online, scholars should explore how motivation is directly linked to statistics cognitions. However, so as not to fragment the field further, models of motivation in online learning should first be tested for efficacy relative to one another. For instance, some research has found support for a goal orientation model of online learning, positing that learners who have high domain specific learning goals are more motivated to learn in blended environments, when they perceived technological features to be an opportunity, as opposed to an obstacle (Klein, Noe, & Wang, 2006). Although there are a variety of models that make various predictions about improving motivation (Deci & Ryan, 2011; Petri & Govern, 2012), it is important to consider other features of motivation, such as how it could be viewed as an individual difference that directly relates to environmental perceptions (Chapter 6).

5. SUMMARY: TOWARD AN EMOTIONALLY ADAPTIVE STATISTICS EDUCATION

If we are striving to develop statistics cognitions, we must be sensitive to the features within a student that may alter their ability to demonstrate their literacy, reasoning, and thinking skills. This Chapter has argued that statistical affect can influence statistics cognition. In particular, we examined how anxiety can arise in a student, but the relationship with anxiety and performance is curvilinear (Keeley et al., 2008), suggesting that performance is optimized when there is just the right amount of anxiety. Achieving this optimal amount of anxiety needs empirical exploration, as identifying ways to subtly manipulate anxiety can convey to students information about urgency of tasks, and can ideally be used to modulate motivation.

Further, we have identified how statistical mood is an important precursor to statistical attitudes, and emphasized the need to assess attitudes throughout the course of the term. Because students tend to have negative attitudes about statistics and using technology (3.2.1.), it is not simply enough to measure attitudes, but to work toward developing specific strategies based on cognitive models that can improve the attitudes that students come to class with.

Although the way to improving affect in an online statistics course could take on a variety of theoretical perspectives, we argue that using a cognitive curriculum (Chapter 4) will improve affective outcomes. For an online course to be successful it must be perceived as useful (Song, Singleton, Hill, & Koh, 2004), be well organized (Romiszowski & Cheng, 1992), and conducive to learning (see Chapter 4; 3.1.2).

Because developing a course on cognitive curriculum requires educators to prove their curricular alignment and gather evidence to refine methods (see Brown, 1992), course features are shaped to promote cognition that may increase student satisfaction with learning. For example, as discussed (Chapter 4; 4.1.2), assessments in a course that measures change in cognitive growth must be formative in nature. Similarly, in order to successfully master the skills associated with such a course, students must become accustomed to a degree of discomfort in uncertainty, and feel a degree of efficacy (Finney & Schraw, 2003) and autonomy (Deci & Ryan, 2011) in the choices they make. Collectively, the structure required to assess cognitive objectives is conducive to motivating students, and improving their affect.

Throughout this Chapter we have outlined why affective features are important when assessing cognitive outcomes, but future research must examine the specific relationships short term high intense emotional states, such as anxiety, share with low intense long term emotional features such as mood. Making such predictions will relay on work aimed at developing and testing models of statistics cognition (Chapter 2), as the neurological foundation of cognitive models - such as the neural educational model of statistics cognition - may allow scholars to test specific predictions.

If stimuli can elicit emotional responses that demonstrate neural activation in the amygdala (Adolphs et al., 2002), and problem solving is representative of increased activity in the posterior cingulate and polar frontal areas (Anderson & Fincham, 2014), then scholars can examine how amygdala activation may or may not influence the ability to solve a problem. Such questions are empirically driven and could allow scholars to develop more nuanced models that can determine how affect can be improved in environments that may be at a disadvantage due to pre-existing attitudes and experiences.

A cognitive statistics course must take a holistic approach to learning, that is why we have shifted focus from a discussion of external features in the environment (Chapter 4) to discussion factors residing inside the student - such as emotions - that can help enhance the quality of education. However, affect is but only one example, as the fields of statistics and online education have made used of a variety of individual differences designed to inform the educator and scholar.

REFERENCES

Abd-El-Fattah, S. M. (2005). The effect of prior experience with computers, statistical self- efficacy, and computer anxiety on students' achievement in an introductory statistics course: A partial least squares path analysis. *International Education Journal*, *5*, 71–79.

Adolphs, R., Baron-Cohen, S., & Tranel, D. (2002). Impaired recognition of social emotions following amygdala damage. *Journal of Cognitive Neuroscience*, *14*(8), 1264–1274. doi:10.1162/089892902760807258 PMID:12495531

Anderson, J. R., Albert, M. V., & Fincham, J. M. (2005). Tracing problem solving in real time: fMRI analysis of the subject-paced Tower of Hanoi. *Journal of Cognitive Neuroscience*, *17*(8), 1261–1274. doi:10.1162/0898929055002427 PMID:16197682

Appelbaum, S. H., & Primmer, B. (1990). An HRx for Computer Anxiety. *Personnel*, *67*, 8–11.

Baddeley, A. (1992). Working memory and conscious awareness. In *Theories of memory* (pp. 11–20). Hillsdale, MI: Lawrence Erlbaum Associates.

Balleine, B. W. (2005). Neural bases of food-seeking: Affect, arousal and reward in corticostriatolimbic circuits. *Physiology & Behavior*, *86*(5), 717–730. doi:10.1016/j.physbeh.2005.08.061 PMID:16257019

Baloğlu, M. (1999). *A comparison of mathematics anxiety and statistics anxiety in relation to general anxiety*. Unpublished manuscript. ERIC Document Reproduction Service No, ED436703.

Baloğlu, M. (2002). Psychometric properties of the statistics anxiety rating scale. *Psychological Reports*, *90*(1), 315–325. doi:10.2466/pr0.2002.90.1.315 PMID:11899004

Bandura, A. (1982). Self-efficacy mechanism in human agency. *The American Psychologist*, *37*(2), 122–147. doi:10.1037/0003-066X.37.2.122

Bandura, A. (1988). Self-efficacy conception of anxiety. *Anxiety Research*, *1*(2), 77–98. doi:10.1080/10615808808248222

Bandura, A. (1997). *Self-efficacy: The exercise of control*. New York, NY: W.H. Freeman and Company.

Bartlett, J. C., & Santrock, J. W. (1979). Affect-dependent episodic memory in young children. *Child Development*, *50*(2), 513–518. doi:10.2307/1129430

Beckers, J. J., & Schmidt, H. G. (2001). The structure of computer anxiety: A six-factor model. *Computers in Human Behavior*, *17*(1), 35–49. doi:10.1016/S0747-5632(00)00036-4

Beckers, J. J., & Schmidt, H. G. (2003). Computer experience and computer anxiety. *Computers in Human Behavior*, *19*(6), 785–797. doi:10.1016/S0747-5632(03)00005-0

Ben-Zvi, D., & Garfield, J. (2008). Introducing the emerging discipline of statistics education. *School Science and Mathematics, 108*, 355-361. doi: , 198610.1111/j.1949- 8594.2008.tb17850.xBlaney

Bless, H., & Fiedler, K. (2006). Mood and the regulation of information processing and behavior. In J. P. Forgas (Ed.), *Affect in social thinking and behavior* (pp. 65–84). New York, NY: Psychology Press.

Blum, K. D. (2005). Gender differences in asynchronous learning in higher education: Learning styles, participation barriers and communication patterns. *Journal of Asynchronous Learning Networks*, *3*, 44–66.

Bower, G. H. (1981). Mood and memory. *The American Psychologist*, *36*(2), 129–148. doi:10.1037/0003-066X.36.2.129 PMID:7224324

Bozionelos, N. (2004). Socio-economic background and computer use: The role of computer anxiety and computer experience in their relationship. *International Journal of Human-Computer Studies*, *61*(5), 725–746. doi:10.1016/j.ijhcs.2004.07.001

Bradley, D. R., & Wygant, C. R. (1998). Male and female differences in anxiety about statistics are not reflected in performance. *Psychological Reports*, *82*(1), 245–246. doi:10.2466/pr0.1998.82.1.245 PMID:9520562

Broos, A. (2005). Gender and information and communication technologies (ICT) anxiety: Male self-assurance and female hesitation. *Cyberpsychology & Behavior*, *8*(1), 21–31. doi:10.1089/cpb.2005.8.21 PMID:15738690

Brosnan, M. J. (1998). The impact of computer anxiety and self-efficacy upon performance. *Journal of Computer Assisted Learning*, *14*(3), 223–234. doi:10.1046/j.1365-2729.1998.143059.x

Brown, A. L. (1992). Design experiments: Theoretical and methodological challenges in creating complex interventions in classroom settings. *Journal of the Learning Sciences*, *2*(2), 141–178. doi:10.1207/s15327809jls0202_2

Brown, S. D., Lent, R. W., & Larkin, K. C. (1989). Self-efficacy as a moderator of scholastic aptitude-academic performance relationships. *Journal of Vocational Behavior*, *35*(1), 64–75. doi:10.1016/0001-8791(89)90048-1

Buche, M. W., Davis, L. R., & Vician, C. (2007). A longitudinal investigation of the effects of computer anxiety on performance in a computing-intensive environment. *Journal of Information Systems Education*, *18*, 415–423.

Cadinu, M., Maass, A., Frigerio, S., Impagliazzo, L., & Latinotti, S. (2003). Stereotype threat: The effect of expectancy on performance. *European Journal of Social Psychology*, *33*(2), 267–285. doi:10.1002/ejsp.145

Caine, R., Centa, D., Doroff, C., Horowitz, J. H., & Wisenbaker, V. (1978). Statistics from whom?. *Teaching Sociology*, *6*(1), 37–46. doi:10.2307/1317164

Cambre, M. A., & Cook, D. L. (1987). Measurement and remediation of computer anxiety. *Educational Technology*, *27*, 15–20.

Carlson, K. A., & Winquist, J. R. (2011). Evaluating an active learning approach to teaching introductory statistics: A classroom workbook approach. *Journal of Statistics Education*, *19*, 1–23. Retrieved from https://ww2.amstat.org/publications/jse/v19n1/carlson.pdf

Carr, S. (2000). As distance education comes of age, the challenge is keeping the students. *The Chronicle of Higher Education*, *46*, 39–41.

Cashin, S. E., & Elmore, P. B. (2005). The Survey of Attitudes Toward Statistics scale: A construct validity study. *Educational and Psychological Measurement*, *65*(3), 509–524. doi:10.1177/0013164404272488

Cavanaugh, C., Gillan, K. J., Kromrey, J., Hess, M., & Blomeyer, R. (2004). *The effects of distance education on K-12 student outcomes: A meta-analysis*. Naperville, IL: Learning Point Associates/North Central Regional Educational Laboratory (NCREL).

Cazan, A. M., Cocoradă, E., & Maican, C. I. (2016). Computer anxiety and attitudes towards the computer and the internet with Romanian high-school and university students. *Computers in Human Behavior*, *55*, 258–267. doi:10.1016/j.chb.2015.09.001

Celik, V., & Yesilyurt, E. (2013). Attitudes to technology, perceived computer self-efficacy and computer anxiety as predictors of computer supported education. *Computers & Education*, *60*(1), 148–158. doi:10.1016/j.compedu.2012.06.008

Chamberlain, J. M., Hillier, J., & Signoretta, P. (2015). Counting better? An examination of the impact of quantitative method teaching on statistical anxiety and confidence. *Active Learning in Higher Education*, *16*(1), 51–66. doi:10.1177/1469787414558983

Chen, K. C., & Jang, S. J. (2010). Motivation in online learning: Testing a model of self- determination theory. *Computers in Human Behavior*, *26*(4), 741–752. doi:10.1016/j.chb.2010.01.011

Chew, P. K., & Dillon, D. B. (2014). Statistics anxiety update: Refining the construct and recommendations for a new research agenda. *Perspectives on Psychological Science*, *9*(2), 196–208. doi:10.1177/1745691613518077 PMID:26173253

Christianson, S. A. (2014). *The handbook of emotion and memory: Research and theory*. New York, NY: Psychology Press.

Chua, S. L., Chen, D. T., & Wong, A. F. (1999). Computer anxiety and its correlates: A meta- analysis. *Computers in Human Behavior*, *15*(5), 609–623. doi:10.1016/S0747-5632(99)00039-4

Chuo, Y. H., Tsai, C. H., Lan, Y. L., & Tsai, C. S. (2011). The effect of organizational support, self efficacy, and computer anxiety on the usage intention of e-learning system in hospital. *African Journal of Business Management*, *5*, 5518–5523. doi:10.5897/AJBM11.725

Clore, G. L., & Schnall, S. (2005). The influence of affect on attitude. In D. Albarracin, B. T. Johnson, & M. P. Zanna (Eds.), *The handbook of attitudes* (pp. 437–489). New York, NY: Psychology Press.

Coan, J. A., & Allen, J. J. (Eds.). (2007). *Handbook of emotion elicitation and assessment*. New York, NY: Oxford University Press.

Conrad, D. L. (2002). Engagement, excitement, anxiety, and fear: Learners experiences of starting an online course. *American Journal of Distance Education*, *16*(4), 205–226. doi:10.1207/S15389286AJDE1604_2

Cronbach, L. J., & Snow, R. E. (1977). *Aptitudes and instructional methods: A handbook for research on interactions*. Oxford, UK: Irvington.

Cruise, R. J., Cash, R. W., & Bolton, D. L. (1985, August). Development and validation of an instrument to measure statistical anxiety. *American Statistical Association Proceedings of the Section on Statistics Educatio*n, *4*, 92-97.

Dai, D. Y., & Sternberg, R. J. (Eds.). (2004). *Motivation, emotion, and cognition: Integrative perspectives on intellectual functioning and development*. Mahwah, NJ: Lawrence Erlbaum Associates.

Dauphinee, T. L., Schau, C., & Stevens, J. J. (1997). Survey of Attitudes Toward Statistics: Factor structure and factorial invariance for women and men. *Structural Equation Modeling: a Multidisciplinary Journal*, *4*(2), 129–141. doi:10.1080/10705519709540066

Davis, M. (1992). The role of the amygdala in fear and anxiety. *Annual Review of Neuroscience*, *15*(1), 353–375. doi:10.1146/annurev.ne.15.030192.002033 PMID:1575447

Deci, E. L., & Ryan, R. M. (2011). Self-determination theory. In P. A. M. Van Lange, A. W. Kruglanski, & E. T. Higgins (Eds.), *Handbook of theories of social psychology*. Thousand Oaks, CA: SAGE Publications.

Dempster, M., & McCorry, N. K. (2009). The role of previous experience and attitudes toward statistics in statistics assessment outcomes among undergraduate psychology students. *Journal of Statistics Education*, *17*, 1–7. Retrieved from http://ww2.amstat.org/publications/jse/v17n2/dempster.pdf PMID:21603584

DeVaney, T. A. (2010). Anxiety and attitude of graduate students in on-campus vs. online statistics courses. *Journal of Statistics Education*, *18*, 1-15. Retrieved from: http://ww2.amstat.org/publications/jse/v18n1/devaney.pdf

Diaz, D. P., & Cartnal, R. B. (1999). Students learning styles in two classes: Online distance learning and equivalent on-campus. *College Teaching*, *47*(4), 130–135. doi:10.1080/87567559909595802

Doronina, O. V. (1995). Fear of computers. *Russian Education & Society*, *37*(2), 10–28. doi:10.2753/RES1060-9393370210

Drennan, J. A., Kennedy, J., & Pisarski, A. (2005). Factors affecting student attitudes toward flexible online learning in management education. *The Journal of Educational Research*, *98*(6), 331–338. doi:10.3200/JOER.98.6.331-338

Duncan, S., & Barrett, L. F. (2007). Affect is a form of cognition: A neurobiological analysis. *Cognition and Emotion*, *21*(6), 1184–1211. doi:10.1080/02699930701437931 PMID:18509504

Dunn, R. S. (2000). Learning styles: Theory, research, and practice. *National Forum of Applied Educational Research Journal, 13*, 3-22.

Dunn, R. S., & Dunn, K. J. (1978). *Teaching students through their individual learning styles: A practical approach*. Upper Saddle River, NJ: Prentice Hall.

Dunn, R. S., & Dunn, K. J. (1992b). *Teaching secondary students through their individual learning styles*. Boston, MA: Allyn & Bacon.

Dunn, R. S., & Dunn, K. J. (1993). Learning Styles/Teaching Styles: Should They...Can They...Be Matched. *Educational Leadership*, *36*, 238–244.

Edwards, J. M., & Trimble, K. (1992). Anxiety, coping and academic performance. *Anxiety, Stress, and Coping*, *5*(4), 337–350. doi:10.1080/10615809208248370

Eich, E., & Metcalfe, J. (1989). Mood dependent memory for internal versus external events. *Journal of Experimental Psychology. Learning, Memory, and Cognition*, *15*(3), 443–445. doi:10.1037/0278-7393.15.3.443

Emmioğlu, E. S. M. A., & Capa-Aydin, Y. (2012). Attitudes and achievement in statistics: A meta- analysis study. *Statistics Education Research Journal*, *11*, 95–102.

Engle, R. W., & Kane, M. J. (2003). Executive attention, working memory capacity, and a two- factor theory of cognitive control. *Psychology of Learning and Motivation*, *44*, 145–199. doi:10.1016/S0079-7421(03)44005-X

Evans, B. (2007). Statistics attitudes, conceptions, and achievement in introduction undergraduate college statistics. *The Mathematics Educator*, *17*, 24–30.

Fabrigar, L. R., Wegener, D. T., MacCallum, R. C., & Strahan, E. J. (1999). Evaluating the use of exploratory factor analysis in psychological research. *Psychological Methods*, *4*(3), 272–299. doi:10.1037/1082-989X.4.3.272

Feinberg, L. B., & Halperin, S. (1978). Affective and cognitive correlates of course performance in introductory statistics. *Journal of Experimental Education*, *46*(4), 11–18. doi:10.1080/00220973.1978.11011637

Finney, S. J., & Schraw, G. (2003). Self-efficacy beliefs in college statistics courses. *Contemporary Educational Psychology*, *28*(2), 161–186. doi:10.1016/S0361-476X(02)00015-2

Fonteyne, L., De Fruyt, F., Dewulf, N., Duyck, W., Erauw, K., Goeminne, K., & Rosseel, Y. et al. (2015). Basic mathematics test predicts statistics achievement and overall first year academic success. *European Journal of Psychology of Education*, *30*(1), 95–118. doi:10.1007/s10212-014-0230-9

Forgas, J. P. (1995). Mood and judgment: The affect infusion model (AIM). *Psychological Bulletin*, *117*(1), 39–96. doi:10.1037/0033-2909.117.1.39 PMID:7870863

Forgas, J. P. (2008). Affect and cognition. *Perspectives on Psychological Science*, *3*(2), 94–101. doi:10.1111/j.1745-6916.2008.00067.x PMID:26158876

Forgas, J. P., & George, J. M. (2001). Affective influences on judgments and behavior in organizations: An information processing perspective. *Organizational Behavior and Human Decision Processes*, *86*(1), 3–34. doi:10.1006/obhd.2001.2971

Frankola, K. (2001). Why online learners drop out. *Workforce*, *80*, 53–59.

Fuller, R. M., Vician, C., & Brown, S. A. (2006). E-learning and individual characteristics: The role of computer anxiety and communication apprehension. *Journal of Computer Information Systems*, *46*, 103–115. doi:10.1080/08874417.2006.11645917

Gal, I. (2002). Adults statistical literacy: Meanings, components, responsibilities. *International Statistical Review*, *70*(1), 1–25. doi:10.1111/j.1751-5823.2002.tb00336.x

Gal, I., Ginsburg, L., & Schau, C. (1997). Monitoring attitudes and beliefs in statistics education. In *The assessment challenge in statistics education* (Vol. 12, pp. 37-51). Amsterdam, Netherlands: IOS Press.

Gardner, R. C., & Lambert, W. E. (1972). *Attitudes and motivation in second-language learning*. Rowley, MA: Newbury House Publishers.

Garfield, J. B. (1993). Teaching statistics using small-group cooperative learning. *Journal of Statistics Education*, *1*, 1–9.

Giner-Sorolla, R. (1999). Affect in attitude. In S. Chaiken, & Y. Trope (Eds.), Dual process theories in social psychology (pp. 441-461). New York, NY: Guildford Press.

Gist, M. E., & Mitchell, T. R. (1992). Self-efficacy: A theoretical analysis of its determinants and malleability. *Academy of Management Review*, *17*, 183–211.

Gross, J. J. (Ed.). (2014). *Handbook of emotion regulation*. New York, NY: Guilford Press.

Hackbarth, G., Grover, V., & Mun, Y. Y. (2003). Computer playfulness and anxiety: Positive and negative mediators of the system experience effect on perceived ease of use. *Information & Management*, *40*(3), 221–232. doi:10.1016/S0378-7206(02)00006 X

Haines, M. E., Norris, M. P., & Kashy, D. A. (1996). The effects of depressed mood on academic performance in college students. *Journal of College Student Development*, *37*, 519–526.

Harting, K., & Erthal, M. J. (2005). History of distance learning. *Information Technology, Learning and Performance Journal*, *23*, 35–44.

Heiligenstein, E., Guenther, G., Hsu, K., & Herman, K. (1996). Depression and academic impairment in college students. *Journal of American College Health, 45*(2), 59–64. doi:10.1080/07448481.1996.9936863 PMID:8908879

Hogg, M. A., Abrams, D., & Martin, G. N. (2010). Affect. In *Psychology* (5th ed.; pp. 647–677). Harlow, UK: Pearson Education.

Howard, G. S., & Smith, R. D. (1986). Computer anxiety in management: Myth or reality?. *Communications of the ACM, 29*(7), 611–615. doi:10.1145/6138.6143

Hsu, M. K., Wang, S. W., & Chiu, K. K. (2009). Computer attitude, statistics anxiety and self- efficacy on statistical software adoption behavior: An empirical study of online MBA learners. *Computers in Human Behavior, 25*(2), 412–420. doi:10.1016/j.chb.2008.10.003

Hubbard, R. (1997). Assessment and the process of learning statistics. *Journal of Statistics Education, 5*, 1–8. Retrieved from http://ww2.amstat.org/publications/jse/v5n1/hubbard.html

Huett, J. B., Kalinowski, K. E., Moller, L., & Huett, K. C. (2008). Improving the motivation and retention of online students through the use of ARCS-based e-mails. *American Journal of Distance Education, 22*(3), 159–176. doi:10.1080/08923640802224451

Hunsley, J. (1987). Cognitive processes in mathematics anxiety and test anxiety: The role of appraisals, internal dialogue, and attributions. *Journal of Educational Psychology, 79*(4), 388–392. doi:10.1037/0022-0663.79.4.388

Igbaria, M., & Iivari, J. (1995). The effects of self-efficacy on computer usage. *Omega, 23*(6), 587–605. doi:10.1016/0305-0483(95)00035-6

Igbaria, M., & Parasuraman, S. (1989). A path analytic study of individual characteristics, computer anxiety and attitudes toward microcomputers. *Journal of Management, 15*(3), 373–388. doi:10.1177/014920638901500302

Isen, A. M. (1987). Positive affect, cognitive processes, and social behavior. *Advances in Experimental Social Psychology, 20*, 203–253. doi:10.1016/S0065-2601(08)60415-3

Jashapara, A., & Tai, W. C. (2006). Understanding the complexity of human characteristics on e- learning systems: An integrated study of dynamic individual differences on user perceptions of ease of use. *Knowledge Management Research & Practice*, *4*(3), 227–239. doi:10.1057/palgrave.kmrp.8500099

Johns, M., Schmader, T., & Martens, A. (2005). Knowing is half the battle teaching stereotype threat as a means of improving womens math performance. *Psychological Science*, *16*(3), 175–179. doi:10.1111/j.0956-7976.2005.00799.x PMID:15733195

Keefe, J. W. (1985). Assessment of learning style variables: The NASSP task force model. *Theory into Practice*, *24*(2), 138–144. doi:10.1080/00405848509543162

Keeley, J., Zayac, R., & Correia, C. (2008). Curvilinear relationships between statistics anxiety and performance among undergraduate students: Evidence for optimal anxiety. *Statistics Education Research Journal*, *7*, 4–15.

Keller, J. M. (1987). Development and use of the ARCS model of instructional design. *Journal of Instructional Development*, *10*(3), 2–10. doi:10.1007/BF02905780

Kernan, M. C., & Howard, G. S. (1990). Computer anxiety and computer attitudes: An investigation of construct and predictive validity issues. *Educational and Psychological Measurement*, *50*(3), 681–690. doi:10.1177/0013164490503026

Kim, J., Lim, J. S., & Bhargava, M. (1998). The role of affect in attitude formation: A classical conditioning approach. *Journal of the Academy of Marketing Science*, *26*(2), 143–152. doi:10.1177/0092070398262005

Klein, H. J., Noe, R. A., & Wang, C. (2006). Motivation to learn and course outcomes: The impact of delivery mode, learning goal orientation, and perceived barriers and enablers. *Personnel Psychology*, *59*(3), 665–702. doi:10.1111/j.1744-6570.2006.00050.x

Knowles, E., & Kerkman, D. (2007). An Investigation of Students Attitude and Motivation toward Online Learning. *InSight: A Collection of Faculty Scholarship, 2*, 70-80.

Kolb, D. A. (1976). *Learning style inventory*. Boston, MA: Hay Group, Hay Resources Direct.

Kolb, D. A. (1999). *Learning style inventory* (3rd ed.). Boston, MA: Hay Group, Hay Resources Direct.

Kreiner, D. S. (2006). A mastery-based approach to teaching statistics online. *International Journal of Instructional Media, 33*, 73–80.

Lalonde, R.N., & Gardner, R.C. (1993). Statistics as a second language? A model for predicting performance in psychology students. *Canadian Journal of Behavioural Science/Revue canadienne des sciences du comportement, 25*, 108-125. doi: 10.1037/h0078792

Lane, R. D., Nadel, L., Allen, J. J. B., & Kaszniak, A. W. (1999). The study of emotion from the perspective of cognitive neuroscience. In *Cognitive neuroscience of emotion* (pp. 3-11). New York, NY: Oxford University Press.

LeDoux, J. (2012). Rethinking the emotional brain. *Neuron, 73*(4), 653–676. doi:10.1016/j.neuron.2012.02.004 PMID:22365542

LeDoux, J. E. (2000). Emotion circuits in the brain. *Annual Review of Neuroscience, 23*(1), 155–184. doi:10.1146/annurev.neuro.23.1.155 PMID:10845062

Leonard, A., & Harris, H. (1979, August). Learning style in a primary care internal medicine residency program. *Archives of Internal Medicine, 139*(8), 872–875. doi:10.1001/archinte.1979.03630450026010 PMID:464701

Lester, D. (2016). Predicting success in psychological statistics courses. *Psychological Reports, 118*(3), 772–777. doi:10.1177/0033294116647687 PMID:27273557

Lewis, M., Haviland-Jones, J. M., & Barrett, L. F. (Eds.). (2010). *Handbook of emotions*. New York, NY: Guilford Press.

Liaw, S. S., Huang, H. M., & Chen, G. D. (2007). Surveying instructor and learner attitudes toward e-learning. *Computers & Education, 49*(4), 1066–1080. doi:10.1016/j.compedu.2006.01.001

Lim, D. H. (2004). Cross cultural differences in online learning motivation. *Educational Media International, 41*(2), 163–173. doi:10.1080/095239804 10001685784

Lynch, R., & Dembo, M. (2004). The relationship between self-regulation and online learning in a blended learning context. *The International Review of Research in Open and Distributed Learning, 5*. Retrieved from: http://www.irrodl.org/index.php/irrodl/article/view/189/271

Macher, D., Paechter, M., Papousek, I., & Ruggeri, K. (2012). Statistics anxiety, trait anxiety, learning behavior, and academic performance. *European Journal of Psychology of Education, 27*(4), 483–498. doi:10.1007/s10212-011-0090-5

Magel, R. (1998). Using cooperative learning in a large introductory statistics class. *Journal of Statistics Education, 6*. Retrieved from: http://ww2.amstat.org/publications/jse/v6n3/magel.html

Mahar, D., Henderson, R., & Deane, F. (1997). The effects of computer anxiety, state anxiety, and computer experience on users performance of computer based tasks. *Personality and Individual Differences, 22*(5), 683–692. doi:10.1016/S0191-8869(96)00260-7

Marcoulides, G. A. (1988). The relationship between computer anxiety and computer achievement. *Journal of Educational Computing Research, 4*(2), 151–158. doi:10.2190/J5N4-24HK-567V-AT6E

Maricutoiu, L. P. (2014). A meta-analysis on the antecedents and consequences of computer anxiety. *Procedia: Social and Behavioral Sciences, 127*, 311–315. doi:10.1016/j.sbspro.2014.03.262

Marks, I. M., & Nesse, R. M. (1994). Fear and fitness: An evolutionary analysis of anxiety disorders. *Ethology and Sociobiology, 15*(5-6), 247–261. doi:10.1016/0162-3095(94)90002-7

Martocchio, J. J. (1994). Effects of conceptions of ability on anxiety, self-efficacy, and learning in training. *The Journal of Applied Psychology, 79*(6), 819–825. doi:10.1037/0021-9010.79.6.819 PMID:7852207

Mayer, R. E. (1992). *Thinking, problem solving, cognition*. New York, NY: W.H. Freeman/Times Books/Henry Holt & Company.

McInerney, V., McInerney, D. M., & Sinclair, K. E. (1994). Student teachers, computer anxiety and computer experience. *Journal of Educational Computing Research, 11*(1), 27–50. doi:10.2190/94D0-B0AF-NLAX-7RYR

Mikels, J. A., Reuter-Lorenz, P. A., Beyer, J. A., & Fredrickson, B. L. (2008). Emotion and working memory: Evidence for domain specific processes for affective maintenance. *Emotion (Washington, D.C.)*, *8*(2), 256–266. doi:10.1037/1528-3542.8.2.256 PMID:18410199

Molton, K. D., Brown, S. D., & Lent, R. W. (1991). Relation of self-efficacy beliefs to academic outcomes: A meta-analytic investigation. *Journal of Counseling Psychology*, *38*(1), 30–38. doi:10.1037/0022-0167.38.1.30

Moore, D. S. (2001). Undergraduate statistics: Looking to the future. *American Statistical Association*, *55*, 1–6.

Mullins, J., & Sabherwal, R. (2014). *How much information is too much? Effects of computer anxiety and self-efficacy.* Thirty Fifth International Conference on Information Systems, Auckland, New Zealand. Retrieved from: http://aisel.aisnet.org/icis2014/proceedings/HumanBehavior/53/

Murtonen, M. (2005). University students research orientations: Do negative attitudes exist toward quantitative methods?. *Scandinavian Journal of Educational Research*, *49*(3), 263–280. doi:10.1080/00313830500109568

Murtonen, M., & Lehtinen, E. (2003). Difficulties experienced by education and sociology students in quantitative methods courses. *Studies in Higher Education*, *28*(2), 171–185. doi:10.1080/0307507032000058064

Musch, J., & Broder, A. (1999). Test anxiety versus academic skills: A comparison of two alternative models for predicting performance in a statistics exam. *The British Journal of Educational Psychology*, *69*(1), 105–116. doi:10.1348/000709999157608 PMID:10230346

Nesse, R. (1998). Emotional disorders in evolutionary perspective. *The British Journal of Medical Psychology*, *71*(4), 397–415. doi:10.1111/j.2044-8341.1998.tb01000.x PMID:9875953

Nesse, R. M. (1990). Evolutionary explanations of emotions. *Human Nature (Hawthorne, N.Y.)*, *1*(3), 261–289. doi:10.1007/BF02733986 PMID:24222085

Neumann, D. L., Hood, M., & Neumann, M. M. (2009). Statistics? You must be joking: The application and evaluation of humor when teaching statistics. *Journal of Statistics Education*, *17*, 1–16. Retrieved from http://ww2.amstat.org/publications/jse/v17n2/neumann.pdf PMID:21603584

Niculescu, A. C., Tempelaar, D., Leppink, J., Dailey-Hebert, A., Segers, M., & Gijselaers, W. (2015). Feelings and performance in the first year at university: Learning-related emotions as predictors of achievement outcomes in mathematics and statistics. *Electronic Journal of Research in Educational Psychology*, *13*(3), 431–462. doi:10.14204/ejrep.37.15012

Noyes, J., & Garland, K. (2006). Explaining students attitudes toward books and computers. *Computers in Human Behavior*, *22*(3), 351–363. doi:10.1016/j.chb.2004.09.004

Nutt, R. M., & Lam, D. (2011). A comparison of mood-dependent memory in bipolar disorder and normal controls. *Clinical Psychology & Psychotherapy*, *18*(5), 379–386. doi:10.1002/cpp.778 PMID:21953908

O'Brien, B. (2002). Online Student Retention: Can It Be Done?. In P. Barker, & S. Rebelsky (Eds.), *Proceedings of World Conference on Educational Multimedia, Hypermedia and Telecommunications* (pp. 1479-1483). Chesapeake, VA: AACE.

Onwuegbuzie, A. J. (1997). Writing a research proposal: The role of library anxiety, statistics anxiety, and composition anxiety. *Library & Information Science Research*, *19*(1), 5–33. doi:10.1016/S0740-8188(97)90003-7

Onwuegbuzie, A. J. (1998). The dimensions of statistics anxiety: A comparison of prevalence rates among mid-southern university students. *Louisiana Educational Research Journal*, *23*, 2340.

Onwuegbuzie, A. J. (1999). Statistics anxiety among African-American graduate students: An affective filter?. *The Journal of Black Psychology*, *25*(2), 189–209. doi:10.1177/0095798499025002004

Onwuegbuzie, A. J. (2000). Statistics anxiety and the role of self-perceptions. *The Journal of Educational Research*, *93*(5), 323–330. doi:10.1080/00220670009598724

Onwuegbuzie, A. J., & Daley, C. E. (1999). Perfectionism and statistics anxiety. *Personality and Individual Differences*, *26*(6), 1089–1102. doi:10.1016/S0191-8869(98)00214-1

Onwuegbuzie, A. J., DaRos, D., & Ryan, J. (1997). The components of statistics anxiety: A phenomenological study. *Focus on Learning Problems in Mathematics*, *19*, 11–35.

Onwuegbuzie, A. J., & Seaman, M. A. (1995). The effect of time constraints and statistics test anxiety on test performance in a statistics course. *Journal of Experimental Education*, *63*(2), 115–124. doi:10.1080/00220973.1995.9943816

Onwuegbuzie, A. J., & Wilson, V. A. (2003). Statistics Anxiety: Nature, etiology, antecedents, effects, and treatmentsa comprehensive review of the literature. *Teaching in Higher Education*, *8*(2), 195–209. doi:10.1080/1356251032000052447

Pajares, F. (1996). Self-efficacy beliefs in academic settings. *Review of Educational Research*, *66*(4), 543–578. doi:10.3102/00346543066004543

Pajares, F. (2003). Self-efficacy beliefs, motivation, and achievement in writing: A review of the literature. *Reading & Writing Quarterly*, *19*(2), 139–158. doi:10.1080/10573560308222

Pajares, F. (2005). Gender differences in mathematics self-efficacy beliefs. In A. M. Gallagher & J. C. Kaufman (Eds.), *Gender differences in mathematics: An integrative psychological approach* (pp. 294–315). New York, NY: Cambridge University Press.

Pashler, H., McDaniel, M., Rohrer, D., & Bjork, R. (2008). Learning styles concepts and evidence. *Psychological Science in the Public Interest*, *9*(3), 105–119. doi:10.1111/j.1539-6053.2009.01038.x PMID:26162104

Pekrun, R. (1992). The impact of emotions on learning and achievement: Towards a theory of cognitive/motivational mediators. *Applied Psychology*, *41*(4), 359–376. doi:10.1111/j.1464-0597.1992.tb00712.x

Pekrun, R., Goetz, T., Titz, W., & Perry, R. P. (2002). Academic emotions in students selfregulated learning and achievement: A program of qualitative and quantitative research. *Educational Psychologist*, *37*(2), 99–105. doi:10.1207/S15326985EP3702_4

Perepiczka, M., Chandler, N., & Becerra, M. (2011). Relationship between graduate students statistics self-efficacy, statistics anxiety, attitude toward statistics, and social support. *Professional Counselor*, *1*(2), 99–108. doi:10.15241/mpa.1.2.99

Perney, J., & Ravid, R. (1990). *The relationship between attitudes toward statistics, math self- concept, test anxiety and graduate students' achievement in an introductory statistics course.* Paper presented at the Annual Meeting of the American Educational Research Association, Boston, MA. ERIC Document Reproduction Service No, ED318607.

Peters, E., Västfjäll, D., Gärling, T., & Slovic, P. (2006). Affect and decision making: A hot topic. *Journal of Behavioral Decision Making*, *19*(2), 79–85. doi:10.1002/bdm.528

Petri, H. L., & Govern, J. M. (2012). *Motivation: Theory, research, and application* (6th ed.). Belmont, CA: Wadsworth Publishing.

Popovich, P. M., Gullekson, N., Morris, S., & Morse, B. (2008). Comparing attitudes towards computer usage by undergraduates from 1986 to 2005. *Computers in Human Behavior*, *24*(3), 986–992. doi:10.1016/j.chb.2007.03.002

Powell, A. L. (2013). Computer anxiety: Comparison of research from the 1990s and 2000s. *Computers in Human Behavior*, *29*(6), 2337–2381. doi:10.1016/j.chb.2013.05.012

Putwain, D. W., Woods, K. A., & Symes, W. (2010). Personal and situational predictors of test anxiety of students in post-compulsory education. *The British Journal of Educational Psychology*, *80*(1), 137–160. doi:10.1348/000709909X466082 PMID:19646333

Ramirez, C., Schau, C., & Emmioglu, E. (2012). The importance of attitudes in statistics education. *Statistics Education Research Journal*, *11*, 57–71.

Riechmann, S. W., & Grasha, A. F. (1974). A rational approach to developing and assessing the construct validity of a student learning style scales instrument. *The Journal of Psychology*, *87*(2), 213–223. doi:10.1080/00223980.1974.9915693

Rodarte-Luna, B., & Sherry, A. (2008). Sex differences in the relation between statistics anxiety and cognitive/learning strategies. *Contemporary Educational Psychology*, *33*(2), 327–344. doi:10.1016/j.cedpsych.2007.03.002

Rogers, H. W. (1902). Legal education in the United States (Reprint: 4079). *The American Law School Review*, *13*, 13–16.

Rolls, E. (2005). What are emotions, why do we have emotions, and what is their computational basis in the brain. In J. Fellous & M. A. Arbib (Eds.), *Who needs emotions? The brain meets the robot* (pp. 117–146). New York, NY: Oxford University Press. doi:10.1093/acprof:oso/9780195166194.003.0005

Romiszowski, A., & Chang, E. C. (1992). Hypertext's contribution to computer-mediated communication: in search of an instructional model. In M. Giardina (Ed.), *Interactive multimedia learning environments* (pp. 111–130). Berlin: Springer. doi:10.1007/978-3-642-77705-9_10

Russell, G., & Bradley, G. (1997). Teachers computer anxiety: Implications for professional development. *Education and Information Technologies*, *2*(1), 17–30. doi:10.1023/A:1018680322904

Russell, J. A. (2003). Core affect and the psychological construction of emotion. *Psychological Review*, *110*(1), 145–172. doi:10.1037/0033-295X.110.1.145 PMID:12529060

Saadé, R. G., Kira, D., & Nebebe, F. (2013). The challenge of motivation in e-Learning: role of anxiety. *Proceedings of the Informing Science and Information Technology Education Conference*, 301-308.

Schacht, S., & Stewart, B. J. (1990). Whats funny about statistics? A technique for reducing student anxiety. *Teaching Sociology*, *18*(1), 52–56. doi:10.2307/1318231

Schmader, T., & Johns, M. (2003). Converging evidence that stereotype threat reduces working memory capacity. *Journal of Personality and Social Psychology*, *85*(3), 440–452. doi:10.1037/0022-3514.85.3.440 PMID:14498781

Schmader, T., Johns, M., & Forbes, C. (2008). An integrated process model of stereotype threat effects on performance. *Psychological Review*, *115*(2), 336–356. doi:10.1037/0033-295X.115.2.336 PMID:18426293

Schunk, D. H., & Ertmer, P. A. (2000). Self-regulation and academic learning: Self-efficacy enhancing interventions. In M., Boekaerts, P.R., Pintrich, & M. Zeidner (Eds.), *Handbook of self-regulation* (pp. 631-649). doi:10.1016/B978-012109890- 2/50048-2

Schunk, D. H., Pintrich, P. R., & Meece, J. R. (2007). *Motivation in education: Theory, research, and applications* (3rd ed.). Upper Saddle River, NJ: Merrill/Prentice Hall.

Schwarz, N., Bless, H., & Bohner, G. (1991). Mood and persuasion: Affective states influence the processing of persuasive communications. *Advances in Experimental Social Psychology, 24*, 161–199. doi:10.1016/S0065-2601(08)60329-9

Schwarz, N., & Clore, G. L. (1983). Mood, misattribution, and judgments of well-being: Informative and directive functions of affective states. *Journal of Personality and Social Psychology, 45*(3), 513–523. doi:10.1037/0022-3514.45.3.513

Seibt, B., & Förster, J. (2004). Stereotype threat and performance: How self-stereotypes influence processing by inducing regulatory foci. *Journal of Personality and Social Psychology, 87*(1), 38–56. doi:10.1037/0022-3514.87.1.38 PMID:15250791

Shell, D. F., Colvin, C., & Bruning, R. H. (1995). Self-efficacy, attributions, and outcome expectancy mechanisms in reading and writing achievement: Grade-level and achievement level differences. *Journal of Educational Psychology, 87*(3), 3867–398. doi:10.1037/0022-0663.87.3.386

Skinner, B. F. (1953). *Science and human behavior*. Chicago, IL: Simon and Schuster.

Smith, C. A., Haynes, K. N., Lazarus, R. S., & Pope, L. K. (1993). In search of the hot cognitions: Attributions, appraisals, and their relation to emotion. *Journal of Personality and Social Psychology, 65*(5), 916–929. doi:10.1037/0022-3514.65.5.916 PMID:8246115

Snow, R. E. (1989). Aptitude-treatment interaction as a framework for research on individual differences in learning. In *Learning and individual differences* (pp. 13-60). New York, NY: W.H. Freeman.

Song, L., Singleton, E. S., Hill, J. R., & Koh, M. H. (2004). Improving online learning: Student perceptions of useful and challenging characteristics. *The Internet and Higher Education, 7*(1), 59–70. doi:10.1016/j.iheduc.2003.11.003

Spencer, S. J., Steele, C. M., & Quinn, D. M. (1999). Stereotype threat and womens math performance. *Journal of Experimental Social Psychology, 35*(1), 4–28. doi:10.1006/jesp.1998.1373

Spielberger, C. D., Gorsuch, R. L., Lushene, R., Vagg, P. R., & Jacobs, G. A. (1983). *Manual for the state-trait anxiety inventory*. Palo Alto, CA: Consulting Psychologists Press.

Steele, C. M. (1997). A threat in the air: How stereotypes shape intellectual identity and performance. *The American Psychologist*, *52*(6), 613–629. doi:10.1037/0003-066X.52.6.613 PMID:9174398

Strecher, V. J., McEvoy DeVellis, B., Becker, M. H., & Rosenstock, I. M. (1986). The role of self- efficacy in achieving health behavior change. *Health Education Quarterly*, *13*(1), 73–92. doi:10.1177/109019818601300108 PMID:3957687

Suanpang, P., Petocz, P., & Kalceff, W. (2004). Student attitudes to learning business statistics: Comparison of online and traditional methods. *Journal of Educational Technology & Society*, *7*, 9–20.

Summers, J. J., Waigandt, A., & Whittaker, T. A. (2005). A comparison of student achievement and satisfaction in an online versus a traditional face-to-face statistics class. *Innovative Higher Education*, *29*(3), 233–250. doi:10.1007/s10755-005-1938-x

Taatgen, N. A., Van Rijn, H., & Anderson, J. (2007). An integrated theory of prospective time interval estimation: The role of cognition, attention, and learning. *Psychological Review*, *114*(3), 577–598. doi:10.1037/0033-295X.114.3.577 PMID:17638497

Tawney, R. H. (1930). The problem of the public schools. *The Political Quarterly*, *14*(2), 117–149. doi:10.1111/j.1467-923X.1943.tb01000.x

Tempelaar, D., & Nijhuis, J. (2007). Commonalities in attitudes and beliefs toward different academic subjects. In *The challenges of educating people to lead in a challenging world* (pp. 225-249). Springer. doi:10.1007/978-1-4020-5612-3_11

Tempelaar, D. T., Gijselaers, W. H., van der Loeff, S. S., & Nijhuis, J. F. (2007). A structural equation model analyzing the relationship of student achievement motivations and personality factors in a range of academic subject-matter areas. *Contemporary Educational Psychology*, *32*(1), 105–131. doi:10.1016/j.cedpsych.2006.10.004

Terrace, H. S. (1963). Discrimination learning with and without errors. *Journal of the Experimental Analysis of Behavior*, *6*(1), 1–27. doi:10.1901/jeab.1963.6-1 PMID:13980667

Terrell, S. R., & Dringus, L. (2000). An investigation of the effect of learning style on student success in an online learning environment. *Journal of Educational Technology Systems*, *28*, 231–238. doi:10.2190/R53M-BVBD-UGV5-77EH

Thatcher, J. B., & Perrewe, P. L. (2002). An empirical examination of individual traits as antecedents to computer anxiety and computer self-efficacy. *Management Information Systems Quarterly*, *26*(4), 381–396. doi:10.2307/4132314

Tice, D. M., Baumeister, R. F., Shmueli, D., & Muraven, M. (2007). Restoring the self: Positive affect helps improve self-regulation following ego depletion. *Journal of Experimental Social Psychology*, *43*(3), 379–384. doi:10.1016/j.jesp.2006.05.007

Tremblay, P.F., Gardner, R.C., & Heipel, G. (2000). A model of the relationships among measures of affect, aptitude, and performance in introductory statistics. *Canadian Journal of Behavioural Science/Revue canadienne des sciences du comportement, 32*, 40-48. doi: 10.1037/h0087099

Tucker, L. R., & Lewis, C. (1973). A reliability coefficient for maximum likelihood factor analysis. *Psychometrika*, *38*(1), 1–10. doi:10.1007/BF02291170

Tudor, G. (2006). Teaching introductory statistics online: Satisfying the students. *Journal of Statistics Education*, *14*, 1–17. Retrieved from www.amstat.org/publications/jse/v14n3/tudor.html

Van Toller, S. (1988). Emotion and the brain. In S. Van Toller & G. H. Dodd (Eds.), *Perfumery: The psychology and biology of fragrance* (pp. 121–143). Springer. doi:10.1007/978-94-009-1215-1_7

Venkatesh, V. (1999). Creation of favorable user perceptions: Exploring the role of intrinsic motivation. *Management Information Systems Quarterly*, *23*(2), 239–260. doi:10.2307/249753

Vogel, T., & Wanke, M. (2016). *Attitudes and attitude change. Social psychology: A modular course* (2nd ed.). New York, NY: Routledge.

Vygotsky, L. S. (1986). *Thought and language (E. Hanfmann & G. Vakar, Trans.)*. Cambridge, MA: MIT Press.

Weingartner, H., Miller, H., & Murphy, D. L. (1977). Mood-state-dependent retrieval of verbal associations. *Journal of Abnormal Psychology, 86*(3), 276–284. doi:10.1037/0021-843X.86.3.276 PMID:874186

Whalen, P. J., Kagan, J., Cook, R. G., Davis, F. C., & Kim, H. (2004). Human amygdala responsivity to masked fearful eye whites. *Science, 306*(5704), 2061–2061. doi:10.1126/science.1103617 PMID:15604401

Whitley, B. E. Jr. (1997). Gender differences in computer-related attitudes and behavior: A meta- analysis. *Computers in Human Behavior, 13*(1), 1–22. doi:10.1016/S0747-5632(96)00026-X

Wilfong, J. D. (2006). Computer anxiety and anger: The impact of computer use, computer experience, and self-efficacy beliefs. *Computers in Human Behavior, 22*(6), 1001–1011. doi:10.1016/j.chb.2004.03.020

Williams, A. S. (2010). Statistics anxiety and instructor immediacy. *Journal of Statistics Education, 18*, 1–18. Retrieved from http://ww2.amstat.org/publications/jse/v18n2/williams.pdf

Williams, P. E., & Hellman, C. M. (2004). Differences in self-regulation for online learning between first-and second-generation college students. *Research in Higher Education, 45*(1), 71–82. doi:10.1023/B:RIHE.0000010047.46814.78

Willingham, D. T., Hughes, E. M., & Dobolyi, D. G. (2015). The scientific status of learning styles theories. *Teaching of Psychology, 42*(3), 266–271. doi:10.1177/0098628315589505

Wine, J. D. (1980). Cognitive-attentional theory of test anxiety. In I.G., Sarason (Ed.), *Test anxiety: Theory, research, and applications* (pp. 349-385). Mahwah, NJ: Lawrence Erlbaum Associates.

Woodrow, J. E. (1992). The influence of programming training on the computer literacy and attitudes of preservice teachers. *Journal of Research on Computing in Education, 25*(2), 200–219. doi:10.1080/08886504.1992.10782044

Wyer, R. S. Jr, Clore, G. L., & Isbell, L. M. (1999). Affect and information processing. *Advances in Experimental Social Psychology, 31*, 1–77. doi:10.1016/S0065-2601(08)60271-3

Zeidner, M. (1991). Statistics and mathematics anxiety in social science students: Some interesting parallels. *The British Journal of Educational Psychology*, *61*(3), 319–328. doi:10.1111/j.2044-8279.1991.tb00989.x PMID:1786211

Ziabreva, I., Poeggel, G., Schnabcl, R., & Braun, K. (2003). Separation-induced receptor changes in the hippocampus and amygdala of Octodon degus: Influence of maternal vocalizations. *The Journal of Neuroscience*, *23*, 5329–5336. PMID:12832558

Chapter 6
Meaningful Individual Differences in Statistics Cognition

ABSTRACT

This chapter focuses on the understanding and use of individual differences in statistics cognition. We argue that individual differences can be classified along a continuum ranging from within an individual (internally derived) to an outside source (externally prescribed), and that where an individual differences falls on the continuum may have important implications for how individual differences are used to describe, control for, predict, or explain findings in scholarly research. We argue that individual differences are more useful when they meaningfully pertain to cognitive development, and outline how motivation (using goal orientation and self-determination theory) can be used as an individual difference. We conclude with a discussion of aligning motivational goals and how online courses could adapt themselves to student motivational profiles.

We already know that even modest amounts of statistical training can have a big impact... and we have little basis for predicting how much more improvement is feasible. - Ziva Kunda and Richard E. Nisbett, 1986, p. 222

DOI: 10.4018/978-1-5225-2420-5.ch006

1. INTRODUCTION

To understand how cognition can be developed in an online statistics classroom, it is necessary to consider how cognitions are quantified (Chapter 2), the influence they have on learning (Chapter 3), their impact on the environment (Chapter 4) and relationship with affect (Chapter 5). However, the fields of statistics (Baloğlu, 2003; Walsh & Ugumba-Agwunobi, 2002) and online education (Muilenburg & Berge, 2005; Van den Branden & Lambert, 1999) commonly examine the role individual differences play in learning, with work linking the basis for such differences to automated cognitive processes (Greenwald, McGhee, & Schwartz, 1998). Such a realization is of great importance for online educators who have the ability to manipulate curriculum to suit individual student needs (see Chapter 4). However, what makes an individual difference meaningful?

In order answer this question, it is first necessary to define the term *individual difference*, and consider how they are used. An individual difference is any identified characteristic that can be used to describe, explain, predict, and control behavior (Coon & Mitterer, 2013). And while many researchers make use of these variables, a need to understand *how* and *why* traditionally used individual differences, such as age (Chyung, 2007), gender (Schram, 1996), socioeconomic status (Bornstein & Bradley, 2014), and learning styles (Yousef, 2016), are theorized to impact general educational outcomes remains (Boekaerts, 1995; Nicholls, Cheung, Lauer, & Patashnick, 1989; Voyer & Voyer, 2014).

This Chapter begins by discussing how individual differences are used in statistics cognition, with special emphasis on their sources (2.1) and how they are used (2.2). We then provide a theoretical framework for selecting more dynamic and meaningful individual differences. By meaningful, we refer to individual differences that can influence cognitive processes (3.1) and consider motivation as an example of an individual difference (3.2), detailing how to determine motivational profiles (3.3) and encourage adaptive motivation in statistical learning (3.4). We then consider how meaningful individual differences may differ when framed in an online context (4), and conclude the Chapter with a summary of what we have learned and where the field can look to go next (5).

2. INDIVIDUAL DIFFERENCES FOR STATISTICS COGNITION

In their simplest form individual differences may be thought of as classification schemes that can be used to explain or controls for differential findings among groups. Yet, as will be discussed, there is more to this process than simply selecting variables that offer a high degree of explanatory power on any particular dataset. To understand the use of individual differences in practice and research, it is first necessary to consider how the theoretical sources of individual differences can influence its application. Once the origin has been examined, a discussion of how individual differences can be used to meet the objectives of modern social sciences is laid out, followed by a review of what types of individuals are most common in statistics education, and are thus likely to be of interest for statistics educators developing courses in online environments.

2.1. Uncovering Sources of Individual Differences

Since the age of Darwinism, scientists have been fascinated with ways to explain diversity (Dall, Houston, & McNamara, 2004). While initially restricted to generalizations, a push to compare differences among closely related sub-species emerged (Lack 1961). Soon after, local environmental adaptations, viewed as psychological traits (Wilson 1998), were widely accepted as existing within specific populations (Endler, 1986). However, despite its acceptance, research focusing on individual differences remained controversial in its scope and focus (Lubinski, 2000). The controversy, in of itself, is nothing new (Mahoney, Buboltz, Levin, Doverspike, Svyantek, 2011), as many of the debates in using individual differences are reminiscent of historic concerns (Ackerman & Heggestad, 1997; Cronbach, 1957; Cronbach & Snow, 1969; Thorndike, 1940).

While these debates are of importance for the nature of social scientific discovery, it may be helpful to classify individual differences based on their theorized sources. After all, if an individual difference is derived from innate capacities, then study is warranted on how this difference can have influenced on student outcomes. Conversely, if the difference is socially construed, then scholars can begin to address modulate any influence they have on learning, while addressing the societal frameworks that allowed such differences to emerge.

2.1.1. Internally Derived Individual Differences

Internally derived individual differences are those that are thought to arise from within an individual. Specifically, individual differences that are operationalized to suggests a genetic foundation. Yet, given this distinction, it may be a false dichotomy to suggest that individual differences are either inherently internal, or external. After all, similar to the previously provided discussion on measuring sources of anxiety (Chapter 5), anxiety can be viewed as arising from a state, but also existing as an internal trait. Although such a distinction may make the classification vague, we maintain that a variety of internally derived variables exist.

For example, most researchers include basic demographic information when conducting research to control for variance related to age, race, intelligence quotient, biological sex, or personality, that are commonly viewed as stable internal constructs that can even be used for making educational predictions. For example, in a systematic meta-analysis of 100 empirical studies, researchers found a significant trend for information processing preferences between older and younger adults (Reed, Chan, & Mikels, 2014). Specifically, older adults had a bias when processing positively valence information, relative to negatively valence information, with an opposite pattern among younger adults. Such a finding is important, as it suggests that as students grow older, their cognitive capacities may be different, not necessarily in terms of ability, but as outlined by Reed and colleagues, at least when it comes to using valenced information to make decisions. As such, age as an internally derived characteristic may hold important meaning for statistics educators.

While the previous example denoted a difference among a cognitive process, age differences have been observed in cognitive components - such as memory - which statistics cognitions are, in part, built upon (Verhaeghen & Salthouse, 1997). For instance, Verhaeghen and Salthouse (1997) noted an age related decline for variables such as episodic memory, or reasoning, suggesting that age could influence a base level cognitive process such as one's memory or ability to reason, that could create a disadvantage in a classroom designed to develop cognition. Similarly, Chyung (2007) noted that in online environments, younger students' self-efficacy increased significantly more than their older counterparts, but that older students interacted significantly more than their younger peers. Importantly, if age could have a direct effect on one's ability to make use of cognitive processes, and limit their growth of self-efficacy (Chapter 5), then older students could be at a disadvantage both in terms of their educational attainment, or find it more difficult to excel on classroom measurements.

Age is just but one difference that educators have used to examine differences (Anastasi, 1937; Salthouse & Babcock, 1991; Tyler, 1947). Collectively, we argue that the purpose of understanding internally sourced individual differences is to recognize how differences can be modulated to enhance educational outcomes. Specifically, internally derived differences cannot be altered by the educator; however, curriculum can be adapted to suit specific student needs. Further, by being aware of possible student differences, educators can modify their environments and assessments to capture cognitive growth in the context of these internal characteristics. Ultimately, recognizing the scientific promise of internally derived individual differences is about considering ways to promote adaptive cognitive responses, while limiting maladaptive cognitive responses.

2.1.2. Externally Prescribed Individual Differences

Externally derived individual differences are those that are theorized to arise from outside an individual. Similar to externally derived individual differences, the theorized source of the difference can, in part, reside within an individual. However, instead of having a genetic basis, these differences are classified in ways that are socially and culturally driven.

Variables such as ethnicity, experience, socioeconomic status, and gender may be best categorized as externally prescribed individual differences. As such, a statistics education based upon cognition would largely consider how environmental differences can shape cognitions, resulting in adaptive or maladaptive learning outcomes. For example, research has long noted that culturally prominent stereotypes can influence academic performance (Cuddy et al., 2009; Steele, 1997). In the traditional paradigm, researchers experimentally vary the presence of a stereotype, and then measure student performance. For instance, in one study, researchers asked American Asian college women to fill out a demographic form and take a math exam (Shih, Pittinsky, & Ambady, 1999). Unknown to participants, the demographic form was designed to highlight a student's gender or ethnic identity. When gender identity was activated, students performed significantly worse; however, when ethnic identity was activated, participants performed significantly better. The authors interpreted these results to suggest that knowledge of stereotypes modulated performance.

The studies conducted by Shih and colleagues (1999) are critical as they highlight central caveats to the use of individual differences. As previously noted, individual differences can be used to describe, explain, predict, or control for behavior. While it may be useful to use gender to describe, explain, or control for variance, it makes less sense to use gender as a predictive variable of future performance, as the nature of the construct is subject to change by culture (Cuddy et al., 2009) and time (Mattern & Schau, 2002). Because gender is not may not be internally derived, it would introduce a great degree of variation in predictive models if used in a uniform fashion. Indeed, even high stakes commonly used assessments, such as the Graduate Record Exam (GRE) are commonly found to be less predictive for success in students with diverse backgrounds (Miller & Stassun, 2014).

However, even if the individual difference is prescribed from an external source, it is important to note that cognitive differences could still be meaningful. We assume that environments can shape cognition (see Chapter 4), and seeing a classroom through a gendered lens could result in meaningful differences for educational outcomes. For instance, teacher expectations can be strong enough to influence anything from stereotypes (Aronson, 2002), to intelligence (Rosenthal & Jacobson, 1968). Influencing who the teacher calls on, and the amount of attention students receive, expectations of others can result in a self-fulfilling prophecy (Gilbert, 1995), that could influence educational attainment.

2.1.3. The Cyclical Relationship Among Sources of Individual Differences

So far, the discussion regarding individual differences has been unilateral in nature. That is, internal (heavily genetic) characteristics can influence generalized cognition, which may or may not manifest in educational outcomes, as having enhancing general cognitions will not result in enhanced use of specialized statistical cognitions (Chapter 3); whereas external (heavily environmental) characteristics can influence domain-specific (Chapter 2) cognitions. Such a distinction has important implications for the neural educational (NE) model of statistics cognition, as it suggests that internal individual differences are first the concern of the researcher, and are always framed in terms of their relationships to neurological components. Yet external individual differences may be more the concern of the educator, and who must begin their course by ensuring that the environment is conducive to developing cognition.

However, just because these individual differences could operate exclusively, does not mean that an interaction is not possible. For example, Borkowski and Ugras (1998), found that older (non-traditional) business students have stronger ethical attitudes relative to their younger peers. In this instance, age is an internal individual difference; however, it would be odd to claim that physical maturation has any direct link to ethical attitudes. Perhaps ethical attitudes are more of a product of society and culture when these older students were growing up? Or perhaps elderly individuals face a stereotype of being kind, and fulfill this expectation? Regardless of the exact reason, an environmental explanation seems more reasonable, suggesting that how this individual is defined by their social and cultural values, systematically varies by age.

Similarly, one can consider how environments can systematically vary by genes, especially in a higher education online learning environment. For instance, scholars have argued that children actively select and create their preferred environments (Scarr, 1992), and that genes actively influence our experiences (Plomin, 1994). Indeed, scholars have even argued that genetics may play a larger role in adulthood, as adults have greater freedom to select their environments (Plomin & Petrill, 1997). As such, it stands to reason that students who enroll in online education classes at the University level will be, to some degree, self-selecting for their preferred learning environments.

Individual differences are complicated because they inherently involve reciprocal relationships with the environment (Manuck & McCaffrey, 2014). So long as internal characteristics are operationalized in the framework of their neurological or genetic basis, they can be of great use to researchers and practitioners looking to identify students who students who could benefit from a curricular adjustment (Chapter 4). Further, by determining what environmental cues are conducive to developing statistics cognitions, educators can ensure that appropriate messages are conveyed to students. However, just as humans develop, so do educational structures. People and structures are both dynamic systems, and as a result, require continuous assessment so as to understand how internal and external features spur or halt development.

2.2. Applications of Individual Differences in Statistics Education

Because individual differences can arise from external or internal sources, researchers and practitioners must be cautious when considering ways to replicate or attenuate their effects in digital statistics classrooms. For instance, if ethical attitudes were genetically sourced (linked to genetics and changed as a product of human development), then Borkowski and Ugras (1998) could have suggested that students below a certain age should be prevented from dealing with ethical dilemmas, as they would not be equipped to handle such tasks. However, because ethical attitude is more likely to be heavily externally prescribed, important insights can be gathered by considering how ethical attitudes develop in classroom.

This example, while abstract, argues for a differential use of individual differences based on their source. However, for individual differences to be an integral part of statistics education, empirical evidence is needed to drive theory. Specifically, within the context of online education, evidence must be gathered, indicating that individual differences have been used to describe, explain, predict, and control for behavior in meaningful ways. As such, the following subsections provide a brief overview of how individual differences have been used in some of the few empirical articles that have examined online statistical environments.

2.2.1. Differences Used to Describe in Statistics Education

The most common way individual differences can be used to describe is when collected for demographic categorization. For example, nearly all studies that examined student outcomes for online statistics courses include information relating to the students gender, age, ethnicity, with fewer studies examining variables such as competency, or affect (Brown & Kulikowich, 2004; Harrington, 1999; Utts, Sommer, Acredolo, Magher, Matthews, 2003). Generally, these variables are used to provide information that helps the reader understand or preemptively alter their sample.

For instance, Harrington (1999) used GPA to measure how competent students were in statistics, whereas others have administered exams scored on a 0 - 5 scale to split students into high and low competency groups (Schuyten & Dekeyser, 1996). While such descriptors may often be overlooked, they are paramount in allowing researchers to generalize findings (Polit & Beck,

2010), that ultimately informs scholars of plausible systematic differences among individual characteristics.

2.2.2. Differences Used to Control for Statistics Education

In addition to describing, scholars have also used a slew of individual differences to control for unsystematic variance. Namely, the idea is, by attributing some of the unexplained variance to individual differences, it becomes easier to detect an effect. However, as will be discussed (3.1.1), although it may be tempting to add additional explanatory variables to predictive models, they must have a strong theoretical basis to warrant inclusion.

Interestingly, many of the variables used to describe are also used to control for. For example, although some scholars have sought to describe competency using individual differences (Brown & Kulikowich, 2004; Harrington, 1999), others have used individual differences to control for differences in competency (Krause, Stark, & Mandl, 2009). Generally, controlling for individual differences is a means to avoid naturally occurring confounds. For instance, one substantial portion of the digital statistics education literature is on the study of educational tools and resources (see Mills & Raju, 2011).

For example, Dutton and Dutton (2005) sought to examine how using a course website that included announcements, lessons, and PowerPoint™ slides would influence educational outcomes. However, directly examining classroom performance may be complicated, as students with higher GPAs may be more likely to do better relative to those with lower GPAs, that could wash out the presence of any effect. Importantly, these researchers were interested in finding out how well students perform after controlling for typical academic performance. Controlling for prior performance is common place (Dierker., Cooper, Alexander, Selya, & Rose, 2015; Dotterweich & Rochelle, 2012), as controlling for measures such as GPA allows scholars to generate a more convincing argument that performance differences actually exists.

2.2.3. Differences Used to Predict Statistics Education

While it may make sense to insert certain variables into our models, using individual differences to control for unsystematic variance is subtractive if scholars neglect to provide a justified rationale. Many scholars likely include variables such as gender, age, or related competency measures as a means to showcase the robustness of their findings. What is more useful to psychologists is a predictive relationship among variables. By predicting outcomes based on individual differences, education can become more effective and efficient, and seriously concerns can be analyzed and addressed.

For instance, research has noted that students with low GPAs tend to perform significantly worse than students with high GPAs in online statistics classrooms (Harrington, 1999). To explain this finding, Harrington suggests that students with higher GPAs may have performed better (and thus learned more) in previous statistics classes, and are more experienced in adapting to an online learning environment. While such a hypothesis is supported by previous literature (Portier & van Buuren, 1995), it is important to consider that this hypothesis was offered post hoc. While no theoretical mechanism was explicitly tested as to why low GPA students performed less well, such information is informative to both scholars and educators. This study stops short of explanation, as no viable mechanism for the effect is empirically examined. Yet importantly, such a prediction may allow education to become more effective and efficient by helping sort students into appropriate classes based on their individual differences (Chapter 4).

Yet, in most cases, it would seem rather absurd to bar someone from taking a online course, simply because they had a low GPA. While it may make sense to offer students with low GPAs reminders about exams, and messages containing the availability of campus resources, a low GPA is only a single predictor. As will be discussed, measures must be taken to ensure the predictors selected are the most parsimonious way to explain findings.

2.2.4. Differences Used to Explain Statistics Education

Being able to make accurate predictions is imperative to any science. While some predictors have withstood the test of time, and continue to be of use to both educators and researchers, predicting something will happen does not provide information regarding the stability the prediction. A prediction, on its own, can pay off, but it is arguably more scientifically satisfying to use individual differences to explain the nature of an effect. As previously discussed (2.2.3), predictions regarding behavior within a dynamic system are subject to both cultural and societal drift. By explaining why something has occurred, one can develop theories about the developmental trajectory of a construct giving them better insight into when and why things occur.

For instance, Hsu, Wang, and Chui (2009) used computer attitude, statistics anxiety, and self-efficacy to examine the use of statistical software among online MBA students. In their study, a model was developed to understand what features could influence behavioral outcomes. The authors argued statistical anxiety could negatively influence behavioral intentions to use statistical software, but that anxiety can also indirectly influence behavioral outcomes by lowering perceived usefulness and perceived ease of use of software. Directly, such findings can help scholars understand why students may choose not to use statistical software, despite its apparent benefits. Indirectly, this study presents a theoretically driven model that attempts to explain the relationship between anxiety and student behavioral outcomes for statistical software.

Similar to other methods of using individual difference, explanatory variables can suffer misuse; putting one's cart (data) before the horse (theory). Just as students learn the differences between nominal, ordinal, interval, and ratio variables, so must educators and researchers learn how to appropriately apply individual differences as explanatory variables. While individual differences, such as gender or age, may be easily recorded, it is important to think of constructs such as anxiety (Chapter 5) or motivation (4.1) can be used to describe, control, predict, and explain psychological phenomena. It is easy to use an individual difference to describe an individual or sample of data, but it becomes much more difficult to attach formal meaning to these variables in an attempt to make a parsimonious model that explains an important effect in a consistently accurate way.

2.3. The Usual Suspects: Commonly Used Individual Differences

In researching this book, over 1,100 peer reviewed articles were sorted. Of those 1,100 + articles, only 47 examined some facet of teaching statistics in an online environment. Examining online statistical environments is a relatively new area of statistics education. However, despite the small amount of research that exists, scholars have made use of a variety of individual differences to provide meaning to their findings.

For example, of the 47 identified articles, 26 were empirical studies that involved some sort of data. As such, each of these studies contains many descriptions of individual differences located at the sample level. Of these studies, many used demographic descriptors to share their sample characteristics. These variables, some were used to control for, were largely gender, age, ethnicity, GPA, previous experiences with online or statistics courses, as well as grade level (Dottwerweich & Rochelle, 2012; Harrington, 1999; Winquist & Carlson, 2014).

However, despite describing and controlling for individual differences, relatively fewer studies have aimed at making predictions using individual differences. Indeed, even meta-analyses examining online statistics education have cited difficulty in assessing the meta-analytic effect of individual differences such as gender due to the low number of studies focusing on computer-assisted instruction at post-secondary institutions (Larwin & Larwin, 2011). However, researchers are beginning to explore such effects.

For instance, a scholar comparing online and face-to-face business statistics courses found that women perform significantly worse in online statistical environments relative to face-to-face learning environments (Flanagan, 2014). An effect was not replicated for men. Although gender may not traditionally be used as an explanatory or predictive variable, Flanagan's study suggests potential differences among gender, a finding strikingly similar to studies on women in mathematics (Gallagher et al., 2000; Hyde, 1981; Hyde & Mertz, 2009; Lindberg, Hyde, Petersen, & Linn, 2010).

However, one can beg the question as to whether or not one can expect to find cognitively presenting gender differences (Hyde, 2005), and if not, whether using such an individual difference for higher order operations (prediction and explanation) is in fact meaningful when framing statistics on cognition. The answer to this questions rests largely on the nature of these theorized differences (3.2), and what empirical evidence has to say.

To shed light on this issue, a meta-analysis comprised of 45 experimental studies, each with a control, examined differences between online and face-to-face statistics classrooms differences (Sosa, Berger, Saw, & Mary, 2011). While the authors noted that statistics classes with some level of computer assistance demonstrated a moderate performance advantage ($d = 0.33$), contextual evidence suggested that this effect could be due to an artifact of experimental methodology. Specifically, the only significant differences were larger effects for studies that provided extra instructional time to treatment groups relative to control, recruited graduate as opposed to undergraduate students, and employed an embedded assessment. Other commonly examined and controlled for variables, such as prior experience, indicated no collective effect. Such results are consistent with prior meta-analytic studies examining prior experience in statistics (Schram, 1996) or online courses (Allen, Bourhis, Burrell, & Mabry, 2002; Cavanaugh, Gillan, Kromrey, Hess, & Blomeyer, 2004) that have suggested a minimal influence of experience.

While there are variables that are empirically supported, many seem to be used simply because they make conceptual sense. This can lead to problems in findings, as without a unified vision of important individual differences, there are several concerns and suggestions that can be made regarding the traditional use of these individual differences.

2.4. Critiques of Using Individual Differences in Statistics Education

Throughout this Chapter, several concerns regarding the use of individual differences have been eluded to. Specifically, these concerns can be categorized into three broad categories: *model deprivation*, *model saturation*, and *generalized validity*. A brief commentary defining each term and discussing why it is important to individual difference research is used to provide a framework to understand how improvements can be made.

Model deprivation consists of a lack of informative variables within a model, preventing the model from being accurately nuanced in the short term. For instance, consider findings from a study (Matthews, 2000) that found a moderate correlation (r = 0.63) between stork sightings and birth rates among 17 European countries (see section 3). In this instance, one variable is highly correlated with another, but they probably do not share a relationship. While the example from is a fun take exemplifying this concern, even the peer review process does not always catch such misuses.

For example, researchers reported a strong relationship between ambient lighting and the prevalence of myopia (short-sightedness), before a child reaches the age of two (Quinn, Shin, Maguire, & Stone, 1999). Findings were interpreted to suggest that using ambient lighting may cause myopia. However, upon failed replications (Gwiazda, Ong, Held, & Thorn, 2000), researchers found that the presence of a nightlight had more to do with the parents poor vision. Genetics of parents with myopia was not controlled for, leading to an spurious conclusion.

In addition to deprived models, a model can be saturated. In this instance, researchers use too many individual differences in their models. While each individual difference is likely to control or explain for a little extra variance, statisticians know the importance of avoiding saturated models (Marker, 2002). For example, if a dataset contains 200 predictors, all of these predictors could explain 100% of the variance. What this results in is a formula that is highly accurate and predictive of that specific dataset. However, this model is likely to be too complex to explain much of anything, and there is no guarantee that it will be predictive of the larger population. Thus, the question becomes finding the right balance between adding enough variables to make efficient and parsimonious predictions, but not so much to saturate the model.

Although this is a difficult task for many fields, it becomes exceedingly difficult for a newly emerging field such as online statistics education. While both statistics and online education literatures have emerged, there is a lack of uniform approaches to teaching statistics online, harming cross generalizations among articles. For example, while meta-analyses may exist comparing online and face-to-face statistics classes (Larwin & Larwin, 2011; Sosa et al., 2011), there is reason to question the extent through that these preliminary findings can be generalized. Much of this book is dedicated to promoting statistics cognition, meaning that everything from the course design to instructional method must be aimed at improving student cognition. As such, only individual differences that appear to influence cognition are of importance. Because cognitive outcomes (e.g., understanding, application) can differ from traditional behavioral outcomes (correctly plugging numbers into a formula), there is a great need to consider what individual differences have meaning to cognitively framed environments.

3. MEANINGFUL INDIVIDUAL DIFFERENCES FOR STATISTICS COGNITION

No matter the intended use (e.g., describe, control, predict, explain), or source (internally derived, externally prescribed), the most important consideration to make when using an individual difference in is whether the difference represents a highly meaningful divergence of individual cognition. That is, by using this variable, how well can we account for differences at a level meaningful to educational outcomes. Importantly, in an environment that is rooted in cognition (Chapter 4), only differential outcomes with a basis in cognitive differences are meaningful.

This point should be clear to statisticians, as it is akin to the adage of confusing correlation with causation. For instance, consider the study by Matthews (2000), which examined the relationship between stork sightings and the birth rate of 17 European countries, finding a correlation of r = 0.62. Although this is considered a large sized correlation in psychological science (Bosco, Aguinis, Singh, & Field, 2015; Cohen, 1988), one could clearly not interpret this data to mean that birthrates were factually influenced by stork sightings. This example provides the key to understanding how individual differences should be used.

For instance, while it may be okay to describe the relationship between stork sightings and birth rates, it would be odd to use this variable to control, predict, or explain much of any immediate value to birth rates. To our knowledge, there are not studies suggesting that storks act as aphrodisiacs, but this point is outside of the author's realm of expertise. So, how does one classify that level of use (describe, control, predict, and explain) any given individual difference is appropriate for? As will be discussed, the answer to this depends on the source, and the valence of the cognitive outcome. In order to understand what individual differences are most appropriate to use, the discussion first comments on what individual differences have primarily been examined in the field of online statistics education. In outlining the controversies and improvements needed in the field of individual differences, the cognitive basis for individual differences is discussed, with an emphasis on the quality of evidence needed to support one's use of the difference in question.

3.1. A Cognitive Basis of Individual Differences

In order to understand the cognitive basis for individual differences, it is necessary to consider the sources of individual differences (2.1). Individual differences - while continuous - can be thought of as primarily internally (age, race, sex etc), or externally derived (experiences, ethnicity, gender etc). Similarly, as outlined in the neural educational model of statistics cognition (Chapter 2) general cognitions are shaped into statistics cognitions, as per the outlined educational objectives (Chapter 3). Importantly, this means that an individual difference that is internally derived, is one that influences statistics cognitions from innate base level differences. Conversely, externally prescribed differences are those that only alter the outcomes assessed by educational objectives.

For example, consider empirical evidence that has been interpreted to suggest a gender difference in online statistics environments (Flanagan, 2014). Because gender may be culturally prescribed, differences at the level of gender are differences that create noise for educational assessments. There is no reason to believe that one's gender should influence true performance on statistical assessments. In fact, evidence has even suggested that the negative impact of a stereotype in mathematics can be modulated by one's gender identity (Davies, Spencer, & Steele, 2005; Schmader, 2002; Shih et al., 1999).

However, a biological sex difference would be a difference recognized at a cognitive level. That is, a difference that is reflective of true performance. Indeed, from a cognitive perspective, research has long affirmed minimal differences among men and women (Hyde, 1981), often finding that people tend to be more similar than they are different (Hyde, 2005). Again, as presented in our discussion on online statistical environments (Chapter 4), before one can use individual differences to predict or explain, a question arises as to how useful the individual difference is as a way to categorize people.

Further, over time, environmental differences may become ingratiated as part of one's identity, and become indistinguishable from internally prescribed differences. If environments can influence one's genetics, it stands to reason that externally derived individual differences could eventually become internalized (2.1.3). Yet in the immediate, such a distinction has important implications for educational solutions. For instance, an externally prescribed difference is cultural and societal, meaning that the only way to overcome the biased difference is to work toward changing societal and cultural expectations. However, this is a lofty task, and as previously discussed (Chapter 3), from a cognitive standpoint stereotypes are nothing more than patterns of automated thinking, that can serve a useful purpose by assisting our navigation of the social world. Conversely, internally derived differences are those that can be accounted for in advance, and used as a basis to develop efficient and effective interventions.

If cognitive individual differences are the only differences useful to a courses built upon a cognitive curriculum, then it suggests that scholars may want to approach more dynamic variables in terms of classifying individual differences. After all, debates about the usefulness of variables such as gender or ethnicity as useful predictors may be little more than shortcuts. While they can be used to provide important insights as to where bias may exist, one is hard pressed to suggest that curriculum should be differentiated on the basis of one's gender or ethnicity. So what variables are cognitively rooted, and dynamic? While there are many plausible variables, throughout the next section the construct of motivation was selected to illustrate how even a construct often viewed to be situational can be used to understand individual differences.

3.2. Motivation as a Meaningful Individual Difference for Statistics Cognition

Systematic methodological frameworks employed by social scientists can largely be traced back to the early twentieth century, when a behaviorist paradigm shift seized psychological sciences (Watson, 1913) as a reaction to the psychoanalyst tradition, that was becoming increasingly criticized for its lack of scientific objectivity in making unfalsifiable claims (Furfey, 1929; Popper, 1963). While the general veracity and fairness of some of these claims is now under greater scrutiny (Notturno & McHugh, 1986; Popper, 1968), scholars trained in the behaviorist tradition were largely concerned with observable behavior. Because behaviorism framed the understanding of psychological effects via stimulus-response chains (Thorndike, 1911), motivation was primarily viewed as the need to achieve a stable internal equilibrium known as *homeostasis* (Young, 1941).

A drive to achieve homeostasis fit into the early behaviorism frameworks, as it integrated the familiar concepts of classical conditioning as a means to explain primary drives, with operant conditioning used to explain secondary drives (Campbell & Krealing, 1953). By reducing drives to types of conditioning, scholars proposed mathematical formulae that could be used to algebraically solve for one's motivation (Hull, 1943). While early attempts to quantify and empirically assess motivation were promising in predicting physiological responses, new frameworks, such as *social learning theory*, proved to be particularly troubling for Hullian proponents (Grusec, 1992). Although scholars attempted to use Hull's theory of drive reduction to explain how people could observe and learn from behaviors (Sears, 1951), inconsistent data seemed to dismiss drive reduction explanations for motivation (Sears, 1963).

Such problems were only the beginning for early motivational theorists, as cognitive variables such as fear and arousal (Marx, 1960), processing (Weiner, 1969), attributions (Ball, 1982), and individual differences (Weiner, 1990) all proved to be essential in explaining motivational processes, often used to explain differences in student outcomes. For example, motivational theory had long maintained that people are motivation by rewards, and that the greater the reward one faces, the more motivated they should be (Marx, 1960). However, researchers noted that rewards could serve to undermine motivation (Greene, Sternberg, Lepper, 1976).

To illustrate how rewards could undermine motivation, consider a field experiment where children were asked to partake in a familiar (drawing), yet novel (with markers), task (Lepper, Greene, & Nisbett, 1973). Experimenters assigned children to one of three conditions: one where children were told they would get a gold star in exchange for drawing (expected reward); one where children were not told of the gold star reward, but given one anyway (unexpected reward); and a control condition (no reward). After a two week follow-up, experimenters conducted a post-observation to measure drawing activity (duration) among students. Lepper and colleagues found that students who received an *expected* reward, drew less than the no award and unexpected reward conditions. These results were interpreted in the context of an *overjustification hypothesis*, that asserted children who received the expected reward rationalized the activity of drawing as being reward contingent, and found little reason to engage in an activity they were not compensated for, whereas those who received an unexpected / no reward were internally (intrinsically) motivated to engage in this activity. Clearly, with such antithetical findings, cognitive accounts of motivation were becoming more important.

However, even though modern motivational theorists largely acknowledge the importance of motivation, it is still largely examined as an outcome (Ames, 1992; Deci, Vallerand, Pelletier, & Ryan, 1991), rather than an individual difference. In one sense, motivation can be viewed as an outcome, as it is likely comprised of online cognitive processes, that use resources to operate (see Graham & Weiner, 1996). However, motivation is not a predominately static construct. Unlikely measuring relatively deterministic variables such as height or weight, motivation is waning both within and between an individual and their environment, much like the tide of an ocean. As such, motivation cannot be directly quantified, even if all appropriate quantitative indicators are provided (Hull, 1943). Like any other dynamic system, motivation must be seen as an updating live process, with predictions based in percentages of probabilities. By considering how motivation is indexed (Schunk, Pintrich, & Meece 2007), and addressing issues of granularity (Pintrich, 2004; Winne & Perry, 2000), scholars may be able to identify or even develop adaptive motivational profiles to improve student learning.

3.2.1. Indexes of Motivation

One of the characteristics of motivation, that has made it elusive to study, is that no single theory of motivation exists. While experimental approaches demand that scholars test theories against one another (Platt, 1964), theories of motivation often serve drastically different purposes, making theoretical comparisons exceedingly difficult.

One of the most direct ways motivation can be understood as a differential construct is by the way it is indexed (Schunk et al., 2007). The issue of motivational indexing can be understood as it applies to both the tasks used to assess motivation, as well as how it is operationalized. Yet, as will be discussed, the manner that motivation is defined has important implications for its application, and the underlying assumptions about its cognitive structure.

For example, one way motivation has been indexed is through an understanding of self-efficacy. Framed in social cognitive theory (Bandura, 1997), this line of work asserts motivation is goal-directed behavior, that is modulated by personal expectations. In other words, motivation is a process that actions are decided based on the expectations of one's performance (Pintrich, 2003; Wigfield, 1994; Wigfield & Eccles, 2000). This conceptualization implies that motivation and performance are inherently linked, as individuals have a natural desire to feel efficacious, and will strive to demonstrate their efficacy. As such, their index of motivation suggests that motivational interventions are, by design, aimed at increasing self-efficacy, and ensuring that students have an educational environment opportunity to demonstrate efficacy is present.

Conversely, motivation can be indexed as persistence, that is commonly defined as the ability to maintain goal driven actions (Schunk et al., 2007). However, depending on the nature of the task used to gauge persistence, a relationship with performance may or may not be warranted. For instance, scholars could argue that the more problems a student solves in a specific amount of time, the more motivated they were to complete their task, as they persisted through a larger amount of problems. However, such as conceptualization of motivation is problematic, as it does not consider competency as a confounding variable (Barrett & Morgan, 1995; Carlton & Winsler, 1998; Elliot & Dweck, 2005), and while some students may have attempted fewer problems, these students may have been motivated to be more accurate, and thus focused on solving individual problems correctly.

While the exact index of motivation can vary (Schunk et al., 2007), there is value in approaching motivation from different indices, as it is a dynamic construct subject to change at both within and between levels. In order to choose the appropriate index as a starting point, one must consider the grain size that their research or educational questions imply.

3.2.2. Motivational Granularity

Due to the numerous ways motivation can be indexed, it is necessary to consider how to organize perspectives of motivational theory so that they can be compared. As mentioned, one way to organize motivation is to consider what is referred to as its granularity, or grain size (Howard-Rose and Winne, 1993; Pintrich, 2004; Winne, Jamieson-Noel, & Muis, 2001). By grain size, scholars attempt to make a distinction between micro and macro level motivational theories. That is, theories designed to explain immediate internal characteristics, relative to features that influence motivation from the external environment.

For example, one of the most dominant approaches studying macro-level motivational frameworks is self-determination theory (SDT: Deci & Ryan, 2008); a motivational model that outlines a process for how externalized motivators can become internalized processes (Deci & Ryan, 2000). In order for a student to be self-determined autonomy, relatedness, and competence must be fostered (Deci & Ryan, 2011). A student may start off as amotivated, but upon receiving external rewards or punishments, they are externally motivated to act. As they complete this task, their motivation may become less external if they focus on the approval they are getting from others as a means to satisfy their ego. If they eventually arrive at a point where they self-endorse the value of the task, they may start to feel internally motivated.

As these personal goals become synthesized with the self, congruence is reached, that leads to interest and enjoyment. However, the key to internalizing motivation rests within one's perceptions of competence, autonomy, and psychological relatedness, each coalescing to bring about an optimal state of educational motivation (Deci & Ryan, 2000). Reaching a state of self-determinism has been shown to improve performance in a variety of academic domains (Black & Deci, 2000; Gagné & Deci, 2005), academic satisfaction (Ntoumanis, 2001), and may be positively related goal-setting and self-regulatory skills (Palmer & Wehmeyer, 2003). However, fostering self-determination is no easy feat, as evidence has indicated that it a developmental process (Sands & Doll, 1996), requiring time and effort to flourish. Ultimately, as educators, it is reasonable to hope students come to class to learn, and by altering an individual's reliance on external rewards, and prompting them to develop their internal motivation, SDT attempts to achieve just this.

However, much of SDT focuses on how educators and employers can enhance self-determination by adjusting environmental stimuli that delivers perceptions of one's competency, autonomy, and or psychological relatedness (Niemiec & Ryan, 2009; Soenens & Vansteenkiste, 2005; Standage, Duda, & Ntoumanis, 2005; Vallerand, Fortier, & Guay, 1997). While SDT certainly deals with cognitive features within the self, it focuses its adaptations and understanding on how these internalized features influence behavior. Conversely, other theories of motivation have shifted the focus from environmental considerations, to individually held goals (Eison, 1979; Elliot & Church, 1997).

One of the most prominent and enduring theories that approaches motivation from an internal structure is goal orientation theory (Elliot, Murayama, & Pekrun, 2011; Hulleman, Schrager, Bodmann, & Harackiewicz, 2010). Goal orientation theory grew, in part, out of motivational research that considered an attributional locus of control and relative performance stability to predict motivational outcomes (Weiner, 1985; 1986). For example, when considering one's performance, an individual could ascribe their performance to internal and stable, suggesting that their performance was the product of their ability. Conversely, another could view their performance as external and unstable, suggesting that performance was due to luck. Similarly, goal orientation shifted these attributional theories to focus on the way individuals approaches or avoided goals, while considering their overarching motivational preference (Dweck, 1986; Dweck & Leggett, 1988).

As noted in Figure 1, goal orientation theory argues that students can approach tasks with an intent on mastering a task, or on performing well.

Figure 1. Classic representation of goal orientation

	Mastery	**Performance**
Approach	**Motivated to learn for sake of knowledge**	**Motivated to demonstrate their ability to others**
Avoidant	**Motivated to avoid task misconceptions**	**Motivated to avoid looking as if you lack ability**

Although more modern approaches tend to drop mastery avoidant components (Elliot et al., 2011), the categories depicted in Figure 1 have been viewed in dichotomous senses (Elliot & Dweck, 1988; Elliot 1999), suggesting that students tend to prefer either mastery or performance goals, and can be either avoidant or approaching in their task orientation. Further, much research has favored emphasizing mastery learning (Hendricks & Payne, 2011), as mastery motivational profiles can lead to higher intrinsic motivation relative to performance approaches (Elliot & Church. 1997).

Additionally, mastery learning has generally been found to be positively related to self-efficacy (Payne, Youngcourt, Beaubien, 2007), metacognition (Schmidt & Ford, 2003), and seeking feedback (VandeWalle, 2003). In a meta-analysis by Payne and colleagues, mastery oriented learners seemed to outperform performance oriented learners across academic tasks. Payne concluded that this is likely due to the self-regulatory skills that mastery orientation promotes. However, as will be discussed (4.2.1.) this is a rather reductionist manner to categorize motivational states, and may denote that dichotomizing mastery and performance approaches is more of an organizational nuance than a factual division of motivational differences.

As suggested in Figure 2, goal orientation is a micro-level motivational grain, as it works within an individual. Unlike SDT, which is driven by environmental changes and pushes an individual to fulfill psychological needs (feelings of competency, autonomy and relatedness), goal orientation is based on internal drives, that pull one's behavior toward a specific focus. Importantly, the grain size of motivation has implications for how these theories are applied to educational practice. As has been discussed, macro granular theories tend to consider environmental shifts, whereas micro granular

Figure 2. Models of motivational granularity and their influence on students

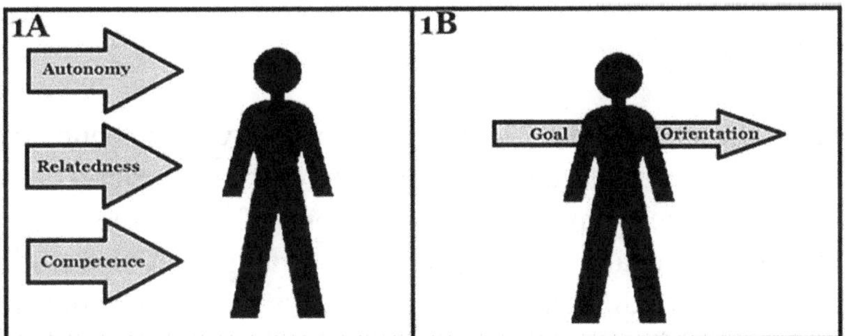

Figure 1A depicts the macro-motivational push of self-determinaiton theory, whereas 1B demonstrates the micro-motivaitonal pull of goal orientation.

theories tend to examine how internal goals direct behaviors. In particular, SDT prompts educators to ask how they can shift amotivation by providing externally motivating punishments and rewards, eventually allowing for motivations to become internalized (Deci & Ryan, 2000). Goal orientation prompts the educator to consider what internal goals students have, and how to best help students fulfill these goals (Midgley, 2014). However, the difference in grain size is more about where the focus starts, as opposed to reflecting on a real distinction between external and internal mechanisms. As will be discussed, micro and macro granular theories share an important joint function.

3.3. Determining Motivational Profiles for Statistics Education

As discussed throughout this section, motivation is a continuous process. It is always updating to reflect one's environment (macro-grain), and subject to individually held goals (micro-grain). While it may be tempting to classify these variables in a dichotomous sense, the world is not often black and white, and just like with externally derived and internally prescribed individual differences (2.1.3.), these features of motivation are likely to share a continuous and reciprocal relationship.

To illustrate the nature of the relationship among micro- and macro-grain motivational influences, discussion will center on the concept of goal orientation alignment, defined as the degree to that a student's goal orientation matches their environment. Then, how the concept of goal orientation alignment can be used to differentiate among motivational profiles will be discussed. Finally, this section will conclude by outlining how motivation fits into the neural educational model, and how it can be used to enhance cognitive processes to create an educationally adaptive motivation.

3.3.1. Goal Orientation Alignment

Goal orientation alignment posits that student motivation is enhanced when students experience relatively alignment among goal orientation in their interactions with their environment or others. That is, the degree that a classroom and an educator's goal orientation is aligned with a student's, the better that student may do. Importantly, this theory asserts that goal orientation is to some degree trait, but is also situational. For example, consider a student whose primary goal is to learn. This student may generally adopt a mastery approach to their learning, but given the right circumstance, such as some sort of competition, a short term goal to perform may override one's enduring mastery goal. Indeed, research has found that students develop a hierarchical structure of their goals (Austin & Vancouver, 1996), with distal (long term goals) considered to be higher ordered.

Further, although goal orientation can be viewed as a state of mind (Elliot & Harackiewicz, 1994), compelling evidence suggests goal orientation can also be viewed as a personal belief (Franken & Brown, 1995; Hertenstein, 2001), mental framework (Ames & Archer, 1987), and even an enduring trait (Bell & Kozlowski, 2002; VandeWalle, 1997). While it may be tempting to adhere to one particular framework, one strength of goal orientation as a divergent theory comes from its ability to provide valuable information at both cognitive and environmental levels (see Deshon & Gillespie, 2005). While there may be an apparent premium on narrowly operationalizing a construct, often scientific advances are driven by the divergence of theoretical frameworks (Kuhn, 1962). As such, goal orientation does not necessarily need to be a state of mind, belief, or trait, as it can include all three features.

Further, if goal orientation can be a state of mind, belief, cognitive framework, and trait, it also relies on cognitive resources, that can wane over time. If motivation is like a muscle, then it stands to reason that even when considering student interests and abilities, individual differences could emerge based on one's innate capacity to replenish motivational resources, or due to situational factors that could drain an individual's motivational resources. As such, goal orientation may not always be predictive of performance. Indeed, some research has supported a preference for mastery orientations (Dweck & Leggett, 1988; Payne et al., 2007), whereas others have found no differential benefits relative to performance orientations (Latham & Marshall, 1982; Latham & Steele, 1983). However, a motivational muscle hypothesis also suggests that micro-grain motivation can dynamically influence macro-grain motivation (and vice versa), in that the more aligned one's individual trait like goal structure is with their environment, the more likely they are to become self-determined within that particular context.

For example, self-determination theory asserts that a student can become self-determined if they have the ability to demonstrate competence, have autonomy, and feel psychologically related (Ryan & Deci, 2002). Importantly, each of these three constructs are more likely to be achieved in an environment where there is total goal alignment. For instance, if a student's goal is to master a skill, they will feel competence when they have been placed in a situation where they were able to develop their skills, and perceive that they are learning for the sake of improving their skills. Conversely, a performance approach student would feel most competent when they are an environment conducive to performing. That is, when placed in a situation where they can demonstrate how smart they are to others, they will feel as though they have demonstrated competence as defined by their goal. Although research has typically framed the development of competence for mastery students (VandeWalle, 1997); it stands to reason that performance approach students would share a desire of demonstrating competence based on their goal.

Similarly, autonomy may also be facilitated via goal orientation alignment. One of the few empirical studies examining personal goals on a construct of self-determination (Garcia & Pintrich, 1996) found no differences in student autonomy. Although Garcia and Pintrich took these results to mean that students do not differ in terms of self-efficacy, test-anxiety, and perceptions of autonomy as a function of intrinsic motivation, they may have a lack of significant results because they divided their sample by orientation (mastery v. performance) as opposed to alignment (match v. mismatch). This is not to suggest that classroom goal orientation influences autonomy differently, but rather the perception of autonomy resides within the student - like their goal orientation - and that a matched goal orientation creates the *perception* of autonomy for the student. For instance, a student who is mastery oriented, and finds themselves well aligned with their educator and classroom, may be likely to believe that they have autonomy because the types of tasks they will be doing, are exactly the type of tasks that the student would like to do. Conversely, a mismatch of goal orientation may lower the student's perception of autonomy, by prompting them to think about how they would rather be doing tasks that allowed them to achieve their personal goals.

Finally, relatedness may also be fostered by goal orientation alignment. If one's personal goal is to achieve, then environments that allow these students to achieve could allow this student to feel as if they belong. Conversely, a mismatch of goal orientation could alienate students, prompting them to think that their classroom, school, or very future in an academic discipline (such as statistics) is not a subject for them!

Goal orientation alignment is important to consider, as it has the potential to allow students to become more self-determined. However, it remains unclear exactly what can or should be done about aligning goal orientation. Grand claims need grand evidence, and while motivational matching may be preferred, no specific prediction is warranted in terms of whether or not performance will increase if goal orientation is aligned. However, it does stand to reason that alignment will result in a higher degree of self-determination, and as discussed, could bring about a variety of positive educational outcomes (Black & Deci, 2000; Gagné & Deci, 2005; Palmer & Wehmeyer, 2003).

3.3.2. Differentiating Motivation

If goal orientation alignment is a key to becoming self-determined, it is reasonable to strive for relative goal orientation alignment. However, just what needs to align, and how much alignment can improve educational outcomes has yet to be seen. For instance, we have previously discussed how curriculum can be differentiated to suit individual needs (see Chapter 4). Similarly, considering to differentiate motivation carries similar concerns. Namely, are motivational profiles and goals so divergent, are separate classes warranted?

From a theoretical perspective, the short answer to this question may be no. That is, student propensity for mastery and performance approaches may not innately differ enough to warrant such a division. While students may have individual characteristics that can influence educational outcomes, researchers have suggested that students commonly pick up on similar clues about the nature of the goals in their learning environments (Deemer, 2004; Meece, Anderman, & Anderman, 2006). Further, as discussed, it is unreasonable to expect that students are always mastery or performance oriented, even within the same task. Some tasks, such as answering one's e-mails, may be too simple to ever really necessitate a mastery goal, whereas other tasks, such as playing the latest video game with a friend, may have less to do with performance and more to do with internal interest, or curiosity.

As a result, it seems reasonable to suggest altering patterns of performance and mastery approach goals within an educational context. If a curriculum is framed on cognitive development and within one's zone of proximal development (ZPD; see Vygotsky, 1986), then students are adequately challenged by their work. Further, by nature of existing within a student's ZPD, the work is by definition adequately difficult. That is, work that can be completed by the student with appropriate educational guidance. If students are adequately challenged by work that has the "right" amount of difficulty, then they can readily work toward mastering the curriculum. However, in order to master the curriculum, students must receive plenty of practice, formative feedback, and reflect on their progress (see Chapter 4). By getting feedback about what can be improved, practicing until they perform better, and reflecting on their progress, students can modulate ineffective learning strategies and expand their knowledge and skills.

However, mastery is not the final variable to the equation. Indeed, because students strive to demonstrate their competence (Ryan & Deci, 2016), they must be given a chance to demonstrate their ability. As such, tasks among the same cognitive learning goal should shift from a mastery to performance orientation once students have mastered the material. In essence, performance oriented tasks will allow the student to demonstrate to themselves, the educator, and maybe even their peers, that the knowledge and skills are those of a master. Further, because the educator has taken care to align their curriculum, students are truly ready to perform, that could attenuate the anxiety for those who felt anxious due to not being educationally prepared (although some students may be predisposed to be anxious; see Chapter 5).

Collectively, such an interpretation suggests that it is unlikely that different classes for different motivational states are necessary above and beyond what is already being offered by gifted and remedial education (Chapter 4). For example if a student is always highly motivated and masters material before their peers, they may become bored with their work and need a greater challenge. Similarly a student who has been unable to master the material, and consistently lags behind peers, may benefit from interventions offered by remedial educators. As a result, the only instance differential classes may be warranted for motivation are when a student continuously masters the material significantly before or after their peers.

3.4. Educationally Adaptive Motivation

By educationally adaptive motivation, we refer to the pattern of motivation that a student may need to succeed in academia. That is, motivation that is optimal to educational outcomes. Because motivation is in a constant state of flux, students need to have the ability to regulate their motivation on academic tasks, based on educational cues (e.g., the amount of points a task is worth implies relative importance and time consumption to the student), and individual drives.

As discussed throughout this section, motivation is both a micro and macro level process, that involves complex interplay among individuals and environments. While important motivational features such as persistence (Dweck, 1986) and grit (Duckworth, Peterson, Matthews, & Kelly, 2007) can be important for educational success, there is evidence to at least suggest different effects based on student motivational profiles, that are comprised of both internal and external factors.

While internal factors such as self-regulatory processes may be readily apparent in how they can influence motivation (Kanfer & Ackerman, 1989), even externally prescribed differences could influence the way motivation functions. For example, research (Mueller & Dweck, 1998) has long suggested that praising one's effort (*mastery*) is better than praising their ability (*performance*). However, others have noted clear gender differences, such as boys preferring ability praise, whereas young girls show a preference for effort praise (Koestner, Zuckerman, and Koestner, 1989). Further, in a recent study, scholars found a prominent gender difference within metaphors about genius (Elmore & Luna-Lucero, 2016), suggesting that men were more likely to be perceived as a genius when their finding was framed as a light bulb "*aha!*" moment of discovery, with women framed as being geniuses when their discoveries were framed as plating a seed, denoting effort and development. In essence, genius for men is framed in a performance sense, whereas it is defined as a mastery process in women. While what makes someone a genius can be debated (Simonton, 1999), at the very least, becoming an expert requires much deliberate practice (Ericsson, 2000), regardless of gender (Ericsson, Prietula., & Cokely, 2007).

As discussed, the primary purpose of understanding motivation as an individual difference is to identify patterns. Specifically, if students share a similar pattern that alternates between mastery and performance based on each student's skill, then educators can externally foster adaptive motivation by adjusting their methods to reflect both mastery and performance practices. Further, if motivation relies on cognitive resources and wanes over time, educationally adaptive motivation becomes largely about setting goals and developing self-regulatory practices (Palmer & Wehmeyer, 2003). While many educators have been advised to help students set goals (Schunk, 1990), it is necessary to consider how students think about these goals in terms of their personal performance and mastery, while helping students identify clear and plausible distal and proximate objectives.

4. ONLINE LEARNING IMPLICATIONS: INDIVIDUAL DIFFERENCES FOR ONLINE STATISTICS COGNITION

While individual differences reside in the student, they come from both internal and external sources. Throughout the literature, individual differences are widely used to help describe, control for, predict, and explain meanings in social science. However, the way we categorize individual differences has consequences for students. We have argued that by framing individual differences as continuous spectrums that explain the transition from trait to state ends we can develop a more realistic model of learning statistics (3.3). Individual differences are merely approximations designed to capture the variance attributed to the data, but instead of framing differences based on often ambiguous markers, that can even change in strength by culture or over time (e.g., ethnicity, age, gender), we need to frame differences on cognitive oriented states.

The implications of using meaningful individual differences for online environments can be considered two folders. First, meaningful individual differences, such as motivation, allow scholars to develop specific predictive models that can be used to assess what types of digital statistical environments are most effective in enhancing cognitions. When scholars develop specific models, they can then provide theories that educators can test in real time. Second, meaningful individual differences should make our results more accurate. Instead of attributing variance to some quasi-independent variable - such as age - we can consider how a variable such as prior experience could provide more immediate insight.

Finally, it is necessary to consider how meaningful individual differences may interact uniquely in online environments. For instance, if meaningful individual differences are both internally and externally sourced, then it means that modifying the external features of a classroom can shape student motivation. Further, by changing the environment, educators are implying different sets of assumptions about existing knowledge base and skill sets. For example, educators probably assume those enrolled in an online class know how to use a computer, check their e-mail, and login to an online web course, but such assumptions may be less apparent in face-to-face environments.

After all, it is not that uncommon for a professor to go an entire semester without a student e-mailing them back. However, we can also turn to empirical research that has implied that different brain regions are activated during mathematical problem solving (Anderson, Pyke, & Fincham, 2016; Chapter 2). One such finding from this research was that the motor cortex activated while students physically solved the problem. However, the motor responses needed for an online course may be inherently different than that of a face-to-face classroom, implying that online environments could require differential activation of the motor cortex relative to face-to-face environments, again causing concern for extrapolating meaning from previously existing studies using course grades without examining the type of curriculum the course was structured upon.

5. SUMMARY: HOW INDIVIDUAL DIFFERENCES CAN MAKE THE DIFFERENCE

There are no easy solutions when it comes to using individual differences. As discussed throughout this Chapter, there is minimal work examining individual differences in online statistics environments. The reviews and meta-analyses that exist (Chance, Ben-Zvi, Garfield, & Medina, 2007; Larwin & Larwin, 2011; Mills & Raju, 2011; Sosa et al., 2011) primarily focus on how technological can be used to enhance student performance, as opposed to discussing differential student outcomes. This is to be expected, for as reviewed in this Chapter and elsewhere (Chapter 4) there may be few performance differences between students in online and face-to-face courses.

While constructs such as age, gender, or ethnicity have helped scholars build models that identify student characteristics that explain variance or predict success (2.3), it is necessary to consider whether these individual differences really say something about the way cognitive processes are developed, or if they are merely meant as shortcuts to detect potential deficits? In other words, scholarly work must begin to address whether differences observed in online learning are due to innate cognitive differences that could require different educational approaches, or are any differences the product of stereotypes and notions that people have about their abilities.

Because of the relative ease curriculum can be differentiated in online environments without student or peer knowledge (Chapter 4), student characteristics could be used to enhance educational experiences by better aligning curriculum to meet student needs if they prove predictive of cognitive differences. However, even if different curriculum could be warranted based on individual differences, it may have more to do with the examples used to deliver content, than the structure of the course or content of the material.

Further, future research must identify and exam the use of meaningful individual differences that allow researchers and educators to make meaningful predictions that can influence the cognitive and affective components of the learning environment. Already scholars are examining the viability of variables such as perspectives of time (Zimbardo & Boyd, 2014), ability to cope (Bonanno & Burton, 2013), and cognitive control (Hendricks & Buchanan, 2016).

We outline a theoretical process that individuals could link student preference for goal orientation to macro-level motivational outcomes (3.2). Such efforts could lead scholars to identify systematic ways to shift the motivational structure of their courses to encourage students who are looking to mastery content, as well as provide a challenge to those who already have. Indeed, one area of statistics education that is generally lacking is that of special statistics education (Chapter 4). Although we have the ability to differentiate tasks and curriculum, we have found no scholarly work that has explicitly examined the cognitive abilities or needs of students with enhanced or damaged cognition.

Individual differences are ways to categorize information that may be of interest. What makes categorizing meaningful, is how well it can predict and explain circumstances in the most parsimonious manner (Gauch, 2003). Yet parsimony can only be achieved through understanding the context behind the data. As suggested by the quote presented at the start of this Chapter, individual differences are much like data - meaningless without context. Whether a scholar, an educator, to use variables appropriately, we must work toward uncovering the context that remains under the individual difference. A good statistician is much like a good detective (Abelson, 2005), as a good statistician must show a level of tenacity in continuously hunting for the best empirical manner to represent objective truth. The consequences of using individual differences in research - especially to predict or explain - are potentially life changing for a student. By understanding where individual differences can arise, and how to select for meaningful differences, even the most dynamic educational processes can be enhanced.

REFERENCES

Abelson. (2005). *Statistics as principled argument*. New York, NY: Psychology Press.

Ackerman, P. L., & Heggestad, E. D. (1997). Intelligence, personality, and interests: Evidence for overlapping traits. *Psychological Bulletin*, *121*(2), 219–245. doi:10.1037/0033-2909.121.2.219 PMID:9100487

Allen, M., Bourhis, J., Burrell, N., & Mabry, E. (2002). Comparing student satisfaction with distance education to traditional classrooms in higher education: A meta-analysis. *American Journal of Distance Education*, *16*(2), 83–97. doi:10.1207/S15389286AJDE1602_3

Ames, C. (1992). Classrooms: Goals, structures, and student motivation. *Journal of Educational Psychology, 84*(3), 261–271. doi:10.1037/0022-0663.84.3.261

Ames, C., & Archer, J. (1987). Mothers beliefs about the role of ability and effort in school learning. *Journal of Educational Psychology, 79*(4), 409–414. doi:10.1037/0022-0663.79.4.409

Anastasi, A. (1937). *Differential psychology*. Oxford, UK: Macmillan.

Anderson, J. R., Pyke, A. A., & Fincham, J. M. (2016). Hidden stages of cognition revealed in patterns of brain activation. *Psychological Science, 27*(9), 1215–1226. doi:10.1177/0956797616654912 PMID:27440808

Aronson, J. (2002). *Stereotype threat: Contending and coping with unnerving expectations. Improving Academic Achievement: Impact of Psychological Factors on Education* (pp. 279–301). San Diego, CA: Academic Press.

Austin, J. T., & Vancouver, J. B. (1996). Goal constructs in psychology: Structure, process, and content. *Psychological Bulletin, 120*(3), 338–375. doi:10.1037/0033-2909.120.3.338

Ball, S. (1982). Motivation. In H. E. Mitzel (Ed.), *Encyclopedia of educational research* (5th ed.). New York, NY: Macmillan.

Baloğlu, M. (2003). Individual differences in statistics anxiety among college students. *Personality and Individual Differences, 34*(5), 855–865. doi:10.1016/S0191-8869(02)00076-4

Bandura, A. (1997). *Self-efficacy: The exercise of control*. New York, NY: W.H. Freeman and Company.

Barrett, K. C., & Morgan, G. A. (1995). Continuities and discontinuities mastery motivation during infancy and toddlerhood: A conceptualization and review. In R. H. MacTurk & G. A. Morgan (Eds.), *Mastery motivation: Origins, conceptualizations and applications* (pp. 57–94). Norwood, NJ: Ablex.

Bell, B. S., & Kozlowski, S. W. (2002). A typology of virtual teams implications for effective leadership. *Group & Organization Management, 27*(1), 14–49. doi:10.1177/1059601102027001003

Black, A. E., & Deci, E. L. (2000). The effects of instructors autonomy support and students autonomous motivation on learning organic chemistry: A self-determination theory perspective. *Science Education*, *84*(6), 740–756. doi:10.1002/1098-237X(200011)84:6<740::AID-SCE4>3.0.CO;2-3

Boekaerts, M. (1995). Self-regulated learning: Bridging the gap between metacognitive and metamotivation theories. *Educational Psychologist*, *30*(4), 195–200. doi:10.1207/s15326985ep3004_4

Borkowski, S. C., & Ugras, Y. J. (1998). Business students and ethics: A meta-analysis. *Journal of Business Ethics*, *17*(11), 1117–1127. doi:10.1023/A:1005748725174

Bornstein, M. H., & Bradley, R. H. (Eds.). (2014). *Socioeconomic status, parenting, and child development*. New York, NY: Routledge.

Bosco, F. A., Aguinis, H., Singh, K., Field, J. G., & Pierce, C. A. (2015). Correlational effect size benchmarks. *The Journal of Applied Psychology*, *100*(2), 431–449. doi:10.1037/a0038047 PMID:25314367

Brown, S. W., & Kulikowich, J. M. (2004). Teaching statistics from a distance: What have we learned?. *International Journal of Instructional Media*, *31*, 19–36.

Campbell, B. A., & Kraeling, D. (1953). Response strength as a function of drive level and amount of drive reduction. *Journal of Experimental Psychology*, *45*(2), 97–101. doi:10.1037/h0060223 PMID:13052836

Carlton, M. P., & Winsler, A. (1998). Fostering intrinsic motivation in early childhood classrooms. *Early Childhood Education Journal*, *25*(3), 159–166. doi:10.1023/A:1025601110383

Cavanaugh, C., Gillan, K. J., Kromrey, J., Hess, M., & Blomeyer, R. (2004). *The effects of distance education on K-12 student outcomes: A meta-analysis*. Naperville, IL: Learning Point Associates/North Central Regional Educational Laboratory (NCREL).

Chance, B., Ben-Zvi, D., Garfield, J. B., & Medina, E. (2007). The Role of Technology in Improving Student Learning of Statistics. *Technology Innovations in Statistics Education Journal, 1*, 1-24. Retrieved from: http://escholarship.org/uc/item/8sd2t4rr

Chyung, S. Y. (2007). Age and gender differences in online behavior, self-efficacy and academic performance. *Quarterly Review of Distance Education*, *8*, 213–222.

Cohen, J. (1988). *Statistical power analysis for the behavioral sciences* (2nd ed.). Hillsdale, NJ: Erlbaum.

Coon, D., & Mitterer, J. O. (2013). *Psychology: A journey*. Belmont, CA: Cengage Learning.

Cronbach, L. J. (1957). The two disciplines of scientific psychology. *The American Psychologist*, *12*(11), 671–684. doi:10.1037/h0043943

Cronbach, L. J., & Snow, R. E. (1969). *Individual differences in learning ability as a function of instructional variables*. Stanford, CA: Stanford University Press.

Cuddy, A. J., Wolf, E. B., Glick, P., Crotty, S., Chong, J., & Norton, M. I. (2015). Men as cultural ideals: Cultural values moderate gender stereotype content. *Journal of Personality and Social Psychology*, *109*(4), 622–635. doi:10.1037/pspi0000027 PMID:26414843

Dall, S. R., Houston, A. I., & McNamara, J. M. (2004). The behavioural ecology of personality: Consistent individual differences from an adaptive perspective. *Ecology Letters*, *7*(8), 734–739. doi:10.1111/j.1461-0248.2004.00618.x

Davies, P. G., Spencer, S. J., & Steele, C. M. (2005). Clearing the air: Identity safety moderates the effects of stereotype threat on womens leadership aspirations. *Journal of Personality and Social Psychology*, *88*(2), 276–287. doi:10.1037/0022-3514.88.2.276 PMID:15841859

Deci, E. L., & Ryan, R. M. (2000). The what and why of goal pursuits: Human needs and the self- determination of behavior. *Psychological Inquiry*, *11*(4), 227–268. doi:10.1207/S15327965PLI1104_01

Deci, E. L., & Ryan, R. M. (2002). *Handbook of self-determination research*. Rochester, NY: University Rochester Press.

Deci, E. L., & Ryan, R. M. (2008). Self-determination theory: A macrotheory of human motivation, development, and health. *Canadian Psychology*, *49*(3), 182–185. doi:10.1037/a0012801

Deci, E. L., & Ryan, R. M. (2011). Self-determination theory. In P. A. M. Van Lange, A. W. Kruglanski, & E. T. Higgins (Eds.), *Handbook of theories of social psychology, 1, 416- 433*. Thousand Oaks, CA: SAGE Publications.

Deci, E. L., Vallerand, R. J., Pelletier, L. G., & Ryan, R. M. (1991). Motivation and education: The self-determination perspective. *Educational Psychologist, 26*(3-4), 325–346. doi:10.1080/00461520.1991.9653137

Deemer, S. (2004). Classroom goal orientation in high school classrooms: Revealing links between teacher beliefs and classroom environments. *Educational Research, 46*(1), 73–90. doi:10.1080/0013188042000178836

DeShon, R. P., & Gillespie, J. Z. (2005). A motivated action theory account of goal orientation. *The Journal of Applied Psychology, 90*(6), 1096–1127. doi:10.1037/0021-9010.90.6.1096 PMID:16316268

Dierker, L., Cooper, J., Alexander, J., Selya, A., & Rose, J. (2015). Evaluating access: A comparison of demographic and disciplinary characteristics of students enrolled in a traditional introductory statistics course vs. a multidisciplinary, project-based course. *Journal of Interdisciplinary Studies in Education, 4*, 22–37.

Dotterweich, D. P., & Rochelle, C. F. (2012). Online, instructional television and traditional delivery: Student characteristics and success factors in business statistics. *American Journal of Business Education, 5*, 129–138.

Duckworth, A. L., Peterson, C., Matthews, M. D., & Kelly, D. R. (2007). Grit: Perseverance and passion for long-term goals. *Journal of Personality and Social Psychology, 92*(6), 1087–1101. doi:10.1037/0022-3514.92.6.1087 PMID:17547490

Dutton, J., & Dutton, M. (2002). How do online students differ from lecture students. *Journal of Asynchronous Learning Networks, 6*, 1–20.

Dweck, C., & Leggett, E. L. (1988). A social-cognitive approach to motivation and personality. *Psychological Review, 95*(2), 256–273. doi:10.1037/0033-295X.95.2.256

Dweck, C. S. (1986). Motivational processes affecting learning. *The American Psychologist, 41*(10), 1040–1048. doi:10.1037/0003-066X.41.10.1040

Eison, J. A. (1979). *The development and validation of a scale to assess different student orientations towards grades and learning* (Unpublished doctoral dissertation). University of Tennessee, Knoxville, TN.

Elliot, A. J. (1999). Approach and avoidance motivation and achievement goals. *Educational Psychologist, 34*(3), 169–189. doi:10.1207/s15326985ep3403_3

Elliot, A. J., & Church, M. A. (1997). A hierarchical model of approach and avoidance achievement motivation. *Journal of Personality and Social Psychology, 72*(1), 218–232. doi:10.1037/0022-3514.72.1.218 PMID:10234849

Elliot, A. J., & Dweck, C. S. (Eds.). (2005). Competence and motivation: Competence as the core of achievement motivation. In *Handbook of competence and motivation*, (pp. 3-12). New York, NY: Guildford Press.

Elliot, A. J., & Harackiewicz, J. M. (1994). Goal setting, achievement orientation, and intrinsic motivation: A mediational analysis. *Journal of Personality and Social Psychology, 66*(5), 968–980. doi:10.1037/0022-3514.66.5.968 PMID:8014838

Elliot, A. J., Murayama, K., & Pekrun, R. (2011). A 3× 2 achievement goal model. *Journal of Educational Psychology, 103*(3), 632–648. doi:10.1037/a0023952

Elliott, E. S., & Dweck, C. S. (1988). Goals: An approach to motivation and achievement. *Journal of Personality and Social Psychology, 54*(1), 5–12. doi:10.1037/0022-3514.54.1.5 PMID:3346808

Elmore, K. C., & Luna-Lucero, M. (2016). Light bulbs or seeds? How metaphors for ideas influence judgments about genius. *Social Psychological & Personality Science, 8*. doi:10.1177/1948550616667611

Endler, J. A. (1986). *Natural selection in the wild*. Princeton, NJ: Princeton University Press.

Ericsson, K. A. (Ed.). (2006). The influence of experience and deliberate practice on the development of superior expert performance. In K.A. Ericsson, N. Charness, R.R. Hoffman, & P.J. Feltovich (Eds.), The *Cambridge handbook of expertise and expert performance*, (vol. 38, pp. 685-705). New York, NY: Cambridge University Press. doi:10.1017/CBO9780511816796.038

Ericsson, K. A., Prietula, M. J., & Cokely, E. T. (2007). The making of an expert. *Harvard Business Review, 85*, 114–121. PMID:17642130

Flanagan, J. L. (2014). Online v. live-taught: Analysis of gender and course format in undergraduate statistics courses in relation to online corporate training. *Academy of Business Journal*, *2*, 89–98.

Franken, R. E., & Brown, D. J. (1995). Why do people like competition? The motivation for winning, putting forth effort, improving ones performance, performing well, being instrumental, and expressing forceful/aggressive behavior. *Personality and Individual Differences*, *19*(2), 175–184. doi:10.1016/0191-8869(95)00035-5

Furfey, P. H. (1929). Psychoanalysis, behaviorism and the gestalt. *Thought*, *4*(2), 237–253. doi:10.5840/thought19294232

Gagné, M., & Deci, E. L. (2005). Self-determination theory and work motivation. *Journal of Organizational Behavior*, *26*(4), 331–362. doi:10.1002/job.322

Gallagher, A. M., Lisi, R. D., Holst, P. C., McGillicuddy-De Lisi, A., Morley, M., & Cahalan, C. (2000). Gender differences in advanced mathematical problem solving. *Journal of Experimental Child Psychology*, *75*(3), 165–190. doi:10.1006/jecp.1999.2532 PMID:10666324

Gauch, H. G. (2003). *Scientific method in practice*. New York, NY: Cambridge University Press.

Gilbert, D. (1995). Attribution and interpersonal perception. In A. I. Tesser (Ed.), *Advanced social psychology* (pp. 98–147). New York, NY: McGraw-Hill.

Graham, S., & Weiner, B. (1996). Theories and principles of motivation. In *Handbook of educational psychology* (Vol. 4, pp. 63-84). New York, NY: Routledge.

Greene, D., Sternberg, B., & Lepper, M. R. (1976). Overjustification in a token economy. *Journal of Personality and Social Psychology*, *34*(6), 1219–1234. doi:10.1037/0022-3514.34.6.1219

Greenwald, A. G., McGhee, D. E., & Schwartz, J. L. (1998). Measuring individual differences in implicit cognition: The implicit association test. *Journal of Personality and Social Psychology*, *74*(6), 1464–1480. doi:10.1037/0022-3514.74.6.1464 PMID:9654756

Grusec, J. E. (1992). Social learning theory and developmental psychology: The legacies of Robert Sears and Albert Bandura. *Developmental Psychology*, *28*(5), 776–786. doi:10.1037/0012-1649.28.5.776

Harrington, D. (1999). Teaching statistics. *Journal of Social Work Education*, *35*, 343–352. doi:10.1080/10437797.1999.10778973

Hendricks, J. W., & Payne, S. C. (2007). Beyond the big five: Leader goal orientation as a predictor of leadership effectiveness. *Human Performance*, *20*, 317–343.

Hendricks, M. A., & Buchanan, T. W. (2016). Individual differences in cognitive control processes and their relationship to emotion regulation. *Cognition and Emotion*, *30*(5), 912–924. doi:10.1080/02699931.2015.1032893 PMID:25947896

Hertenstein, E. J. (2001). Goal orientation and practice condition as predictors of training results. *Human Resource Development Quarterly*, *12*(4), 403–419. doi:10.1002/hrdq.1005

Howard-Rose, D., & Winne, P. H. (1993). Measuring component and sets of cognitive processes in self-regulated learning. *Journal of Educational Psychology*, *85*(4), 591–604. doi:10.1037/0022-0663.85.4.591

Hsu, M. K., Wang, S. W., & Chiu, K. K. (2009). Computer attitude, statistics anxiety and self- efficacy on statistical software adoption behavior: An empirical study of online MBA learners. *Computers in Human Behavior*, *25*(2), 412–420. doi:10.1016/j.chb.2008.10.003

Hull, C. (1943). *Principles of behavior*. New York, NY: Appleton-Century-Crofts, Inc.

Hulleman, C. S., Schrager, S. M., Bodmann, S. M., & Harackiewicz, J. M. (2010). A meta-analytic review of achievement goal measures: Different labels for the same constructs or different constructs with similar labels?. *Psychological Bulletin*, *136*(3), 422–449. doi:10.1037/a0018947 PMID:20438145

Hyde, J. S. (1981). How large are cognitive gender differences? A meta-analysis using! w^2 and d. *The American Psychologist*, *36*(8), 892–901. doi:10.1037/0003-066X.36.8.892

Hyde, J. S. (2005). The gender similarities hypothesis. *The American Psychologist*, *60*(6), 581–592. doi:10.1037/0003-066X.60.6.581 PMID:16173891

Hyde, J. S., & Mertz, J. E. (2009). Gender, culture, and mathematics performance. *Proceedings of the National Academy of Sciences of the United States of America*, *106*(22), 8801–8807. doi:10.1073/pnas.0901265106 PMID:19487665

Kanfer, R., & Ackerman, P. L. (1989). Motivation and cognitive abilities: An integrative/aptitude-treatment interaction approach to skill acquisition. *The Journal of Applied Psychology*, *74*(4), 657–690. doi:10.1037/0021-9010.74.4.657

Koestner, R., Zuckerman, M., & Koestner, J. (1989). Attributional focus of praise and childrens intrinsic motivation: The moderating role of gender. *Personality and Social Psychology Bulletin*, *15*, 61–72. doi:10.1177/0146167289151006

Krause, U. M., Stark, R., & Mandl, H. (2009). The effects of cooperative learning and feedback on e-learning in statistics. *Learning and Instruction*, *19*(2), 158–170. doi:10.1016/j.learninstruc.2008.03.003

Kuhn, T. S. (1962). *The structure of scientific revolutions*. Chicago, IL: University of Chicago Press.

Kunda, Z., & Nisbett, R. E. (1986). Prediction and the partial understanding of the law of large numbers. *Journal of Experimental Social Psychology*, *22*(4), 339–354. doi:10.1016/0022-1031(86)90019-3

Lack, D. (1961). *Darwin's finches*. Cambridge, UK: Cambridge University Press.

Larwin, K., & Larwin, D. (2011). A meta-analysis examining the impact of computer-assisted instruction on postsecondary statistics education: 40 years of research. *Journal of Research on Technology in Education*, *43*(3), 253–278. doi:10.1080/15391523.2011.10782572

Latham, G. P., & Marshall, H. A. (1982). The effect of self-set, participatively set and assigned goals on the performance of government employees. *Personnel Psychology*, *35*(2), 399–404. doi:10.1111/j.1744-6570.1982.tb02204.x

Latham, G. P., & Steele, T. P. (1983). The motivational effects of participation versus goal setting on performance. *Academy of Management Journal*, *26*(3), 406–417. doi:10.2307/256253

Lepper, M. R., Greene, D., & Nisbett, R. E. (1973). Undermining childrens intrinsic interest with extrinsic reward: A test of the overjustification hypothesis. *Journal of Personality and Social Psychology*, *28*(1), 129–137. doi:10.1037/h0035519

Lindbcrg, S. M., Hyde, J. S., Petersen, J. L., & Linn, M. C. (2010). New trends in gender and mathematics performance: A meta-analysis. *Psychological Bulletin*, *136*(6), 1123–1135. doi:10.1037/a0021276 PMID:21038941

Lubinski, D. (2000). Scientific and social significance of assessing individual differences: Sinking shafts at a few critical points. *Annual Review of Psychology*, *51*(1), 405–444. doi:10.1146/annurev.psych.51.1.405 PMID:10751977

Mahoney, K. T., Buboltz, W., Levin, I. P., Doverspike, D., & Svyantek, D. J. (2011). Individual differences in a within-subjects risky-choice framing study. *Personality and Individual Differences*, *51*(3), 248–257. doi:10.1016/j.paid.2010.03.035

Manuck, S. B., & McCaffery, J. M. (2014). Gene-environment interaction. *Annual Review of Psychology*, *65*(1), 41–70. doi:10.1146/annurev-psych-010213-115100 PMID:24405358

Marker, D. (2002). *Model theory: An introduction*. Chicago, IL: Springer.

Marx, M. (1960). Motivation. In L. W. Harris (Ed.), *Encyclopedia of educational research* (3rd ed., pp. 888–901). New York, NY: Macmillan.

Mattern, N., & Schau, C. (2002). Gender differences in science attitude-achievement relationships over time among white middle-school students. *Journal of Research in Science Teaching*, *39*(4), 324–340. doi:10.1002/tea.10024

Matthews, R. (2000). Storks deliver babies (p = 0.008). *Teaching Statistics*, *22*(2), 36–38. doi:10.1111/1467-9639.00013

Meece, J. L., Anderman, E. M., & Anderman, L. H. (2005). Classroom goal structure, student motivation, and academic achievement. *Annual Review of Psychology*, *57*(1), 487–503. doi:10.1146/annurev.psych.56.091103.070258 PMID:16318604

Midgley, C. (Ed.). (2014). *Goals, goal structures, and patterns of adaptive learning*. New York, NY: Routledge.

Miller, C., & Stassun, K. (2014). A test that fails. *Nature*, *510*(7504), 303–304. doi:10.1038/nj7504-303a

Mills, J. D., & Raju, D. (2011). Teaching statistics online: A decade's review of the literature about what works. *Journal of Statistics Education*, *19*, 1–28. Retrieved from http://ww2.amstat.org/publications/jse/v19n2/mills.pdf

Mueller, C. M., & Dweck, C. S. (1998). Intelligence praise can undermine motivation and performance. *Journal of Personality and Social Psychology*, *75*, 33–52. doi:10.1037/0022-3514.75.1.33 PMID:9686450

Muilenburg, L. Y., & Berge, Z. L. (2005). Student barriers to online learning: A factor analytic study. *Distance Education*, *26*(1), 29–48. doi:10.1080/01587910500081269

Nicholls, J. G., Cheung, P. C., Lauer, J., & Patashnick, M. (1989). Individual differences in academic motivation: Perceived ability, goals, beliefs, and values. *Learning and Individual Differences*, *1*(1), 63–84. doi:10.1016/1041-6080(89)90010-1

Niemiec, C. P., & Ryan, R. M. (2009). Autonomy, competence, and relatedness in the classroom: Applying self-determination theory to educational practice. *School Field*, *7*, 133–144. doi:10.1177/1477878509104318

Ntoumanis, N. (2001). A self-determination approach to the understanding of motivation in physical education. *The British Journal of Educational Psychology*, *71*(2), 225–242. doi:10.1348/000709901158497 PMID:11449934

Palmer, S. B., & Wehmeyer, M. L. (2003). Promoting self-determination in early elementary school teaching self-regulated problem-solving and goal-setting skills. *Remedial and Special Education*, *24*(2), 115–126. doi:10.1177/07419325030240020601

Payne, S. C., Youngcourt, S. S., & Beaubien, J. M. (2007). A meta-analytic examination of the goal orientation nomological net. *The Journal of Applied Psychology*, *92*(1), 128–150. doi:10.1037/0021-9010.92.1.128 PMID:17227156

Pintrich, P. R. (2003). A motivational science perspective on the role of student motivation in learning and teaching contexts. *Journal of Educational Psychology*, *95*(4), 667–686. doi:10.1037/0022-0663.95.4.667

Pintrich, P. R. (2004). A conceptual framework for assessing motivation and self-regulated learning in college students. *Educational Psychology Review*, *16*(4), 385–407. doi:10.1007/s10648-004-0006-x

Pittinsky, T. L., Shih, M., & Ambady, N. (1999). Identity adaptiveness: Affect across multiple identities. *The Journal of Social Issues*, *55*(3), 503–518. doi:10.1111/0022-4537.00130

Platt, J. R. (1964). Strong inference. *Science*, *146*(3642), 347–353. doi:10.1126/science.146.3642.347 PMID:17739513

Plomin, R. (1994). *Genetics and experience: The interplay between nature and nurture*. Thousand Oaks, CA: Sage Publications.

Plomin, R., & Petrill, S. A. (1997). Genetics and intelligence: What's new?. *Intelligence*, *24*(1), 53–77. doi:10.1016/S0160-2896(97)90013-1

Polit, D. F., & Beck, C. T. (2010). Generalization in quantitative and qualitative research: Myths and strategies. *International Journal of Nursing Studies*, *47*(11), 1451–1458. doi:10.1016/j.ijnurstu.2010.06.004 PMID:20598692

Popper, K. (1968). *Conjectures and refutations*. New York, NY: Routledge.

Portier, S. J., & van Buuren, H. A. (1995). An interactive learning environment (ILE) to study statistics: Effects of prior knowledge on the use of embedded support devices. *European Journal of Psychology of Education*, *10*(2), 197–207. doi:10.1007/BF03172916

Quinn, G. E., Shin, C. H., Maguire, M. G., & Stone, R. A. (1999). Myopia and ambient lighting at night. *Nature*, *399*(6732), 113–114. doi:10.1038/20094 PMID:10335839

Reed, A. E., Chan, L., & Mikels, J. A. (2014). Meta-analysis of the age-related positivity effect: Age differences in preferences for positive over negative information. *Psychology and Aging*, *29*(1), 1–15. doi:10.1037/a0035194 PMID:24660792

Rosenthal, R., & Jacobson, L. F. (1968). Teacher expectations for the disadvantaged. *Scientific American*, *218*(4), 19–23. doi:10.1038/scientificamerican0468-19

Ryan, R. M., & Deci, E. L. (2016). Facilitating and hindering motivation, learning, and well-being in schools: Research and observations from self-determination theory. In K. R. Wentzel & D. B. Miele (Eds.), *Handbook on motivation at schools* (pp. 96–119). New York, NY: Routledge.

Salthouse, T. A., & Babcock, R. L. (1991). Decomposing adult age differences in working memory. *Developmental Psychology*, *27*(5), 763–776. doi:10.1037/0012-1649.27.5.763

Sands, D. J., & Doll, B. (1996). Fostering self-determination is a developmental task. *The Journal of Special Education*, *30*(1), 58–76. doi:10.1177/002246699603000104

Scarr, S. (1992). Developmental theories for the 1990s: Development and individual differences. *Child Development*, *63*(1), 1–19. doi:10.2307/1130897 PMID:1343618

Schmader, T. (2002). Gender identification moderates stereotype threat effects on womens math performance. *Journal of Experimental Social Psychology*, *38*(2), 194–201. doi:10.1006/jesp.2001.1500

Schmidt, A. M., & Ford, J. K. (2003). Learning within a learner control training environment: The interactive effects of goal orientation and metacognitive instruction on learning outcomes. *Personnel Psychology*, *56*(2), 405–429. doi:10.1111/j.1744-6570.2003.tb00156.x

Schram, C. M. (1996). A meta-analysis of gender differences in applied statistics achievement. *Journal of Educational and Behavioral Statistics*, *21*(1), 55–70. doi:10.3102/10769986021001055

Schunk, D. H. (1990). Goal setting and self-efficacy during self-regulated learning. *Educational Psychologist*, *25*(1), 71–86. doi:10.1207/s15326985ep2501_6

Schunk, D. H., Pintrich, P. R., & Meece, J. R. (2007). *Motivation in education: Theory, research, and applications* (3rd ed.). Upper Saddle River, NJ: Merrill/Prentice Hall.

Schuyten, G., & Dekeyser, H. (1996). Computer software in Statistics Education: Student views on the impact of a computer package on affection and cognition. In *Research on the role of technology in teaching and learning statistics* (pp. 207-217). Voorburg, The Netherlands: ISI.

Sears, R. R. (1951). A theoretical framework for personality and social behavior. *The American Psychologist*, *6*(9), 476–482. doi:10.1037/h0063671

Sears, R. R. (1963). Dependency motivation. In M. R. Jones (Ed.), *Nebraska symposium on motivation* (pp. 25–64). Oxford, UK: University of Nebraska Prcss.

Simonton, D. K. (1999). *Origins of genius: Darwinian perspectives on creativity*. New York, NY: Oxford University Press.

Soenens, B., & Vansteenkiste, M. (2005). Antecedents and outcomes of self-determination in 3 life domains: The role of parents and teachers autonomy support. *Journal of Youth and Adolescence*, *34*(6), 589–604. doi:10.1007/s10964-005-8948-y

Sosa, G. W., Berger, D. E., Saw, A. T., & Mary, J. C. (2011). Effectiveness of computer-assisted instruction in statistics a meta-analysis. *Review of Educational Research*, *81*(1), 97–128. doi:10.3102/0034654310378174

Standage, M., Duda, J. L., & Ntoumanis, N. (2005). A test of self-determination theory in school physical education. *The British Journal of Educational Psychology*, *75*(3), 411–433. doi:10.1348/000709904X22359 PMID:16238874

Steele, C. M. (1997). A threat in the air: How stereotypes shape intellectual identity and performance. *The American Psychologist*, *52*(6), 613–629. doi:10.1037/0003-066X.52.6.613 PMID:9174398

Thorndike, E. L. (1911). *Animal intelligence: Experimental studies*. New York, NY: Macmillan. doi:10.5962/bhl.title.55072

Thorndike, E. L. (1940). *Human nature and the social order*. Oxford, UK: Macmillan. doi:10.1037/14654-000

Tyler, L. E. (1947). *The psychology of human differences*. New York, NY: Appleton-Century Company. doi:10.1037/11271-000

Utts, J., Sommer, B., Acredolo, C., Maher, M. W., & Matthews, H. R. (2003). A study comparing traditional and hybrid internet-based instruction in introductory statistics classes. *Journal of Statistics Education*, *11*, 171–173. Retrieved from https://ww2.amstat.org/publications/jse/v11n3/utts.html

Vallerand, R. J., Fortier, M. S., & Guay, F. (1997). Self-determination and persistence in a real- life setting: Toward a motivational model of high school dropout. *Journal of Personality and Social Psychology*, *72*(5), 1161–1176. doi:10.1037/0022-3514.72.5.1161 PMID:9150590

Van den Branden, J., & Lambert, J. (1999). Cultural issues related to transnational open and distance learning in universities: A European problem?. *British Journal of Educational Technology*, *30*(3), 251–260. doi:10.1111/1467-8535.00114

VandeWalle, D. (1997). Development and validation of a work domain goal orientation instrument. *Educational and Psychological Measurement*, *57*(6), 995–1015. doi:10.1177/0013164497057006009

VandeWalle, D. (2003). A goal orientation model of feedback-seeking behavior. *Human Resource Management Review*, *13*(4), 581–604. doi:10.1016/j.hrmr.2003.11.004

Verhaeghen, P., & Salthouse, T. A. (1997). Meta-analyses of age–cognition relations in adulthood: Estimates of linear and nonlinear age effects and structural models. *Psychological Bulletin*, *122*(3), 231–249. doi:10.1037/0033-2909.122.3.231 PMID:9354147

Voyer, D., & Voyer, S. D. (2014). Gender differences in scholastic achievement: A meta- analysis. *Psychological Bulletin*, *140*(4), 1174–1204. doi:10.1037/a0036620 PMID:24773502

Vygotsky, L. S. (1986). *Thought and language (E. Hanfmann & G. Vakar, Trans.)*. Cambridge, MA: MIT Press.

Walsh, J. J., & Ugumba-Agwunobi, G. (2002). Individual differences in statistics anxiety: The roles of perfectionism, procrastination and trait anxiety. *Personality and Individual Differences*, *33*(2), 239–251. doi:10.1016/S0191-8869(01)00148-9

Watson, J. B. (1913). Psychology as the behaviorist views it. *Psychological Review*, *20*(2), 158–177. doi:10.1037/h0074428

Weiner, B. (1969). Motivation. In R. L. Ebel (Ed.), *Encyclopedia of educational research* (4th ed., pp. 878–888). New York, NY: Macmillan.

Weiner, B. (1985). An attributional theory of achievement motivation and emotion. *Psychological Review*, *92*(4), 548–573. doi:10.1037/0033-295X.92.4.548 PMID:3903815

Weiner, B. (1986). *An attribution theory of motivation and emotion*. New York, NY: Springer- Verlag. doi:10.1007/978-1-4612-4948-1

Weiner, B. (1990). History of motivational research in education. *Journal of Educational Psychology*, *82*(4), 616–622. doi:10.1037/0022-0663.82.4.616

Wigfield, A. (1994). Expectancy-value theory of achievement motivation: A developmental perspective. *Educational Psychology Review*, *6*(1), 49–78. doi:10.1007/BF02209024

Wigfield, A., & Eccles, J. S. (2000). Expectancy–value theory of achievement motivation. *Contemporary Educational Psychology*, *25*(1), 68–81. doi:10.1006/ceps.1999.1015 PMID:10620382

Wilson, D. S. (1998). Adaptive individual differences within single populations. *Philosophical Transactions of the Royal Society of London. Series B, Biological Sciences*, *353*(1366), 199–205. doi:10.1098/rstb.1998.0202

Winne, P. H., & Jamieson-Noel, D. (2003). Self-regulating studying by objectives for learning: Students reports compared to a model. *Contemporary Educational Psychology*, *28*(3), 259–276. doi:10.1016/S0361-476X(02)00041-3

Winne, P. H., & Perry, N. E. (2000). Measuring self-regulated learning. In M. Boekaert, P. R. P. Pintrich, & M. Zeider (Eds.), *Handbook of self-regulation* (pp. 531–566). San Diego, CA: Academic Press. doi:10.1016/B978-012109890-2/50045-7

Winquist, J. R., & Carlson, K. A. (2014). Flipped statistics class results: Better performance than lecture over one year later. *Journal of Statistics Education*, *22*, 1–10. Retrieved from http://ww2.amstat.org/publications/jse/v22n3/winquist.pdf

Young, P. T. (1941). Motivation. In W. S. Monroe (Ed.), *Encyclopedia of educational research* (pp. 735–742). New York, NY: Macmillan.

Yousef, D. A. (2016). Learning styles preferences of statistics students: A study in the faculty of business and economics at the UAE University. *Quality Assurance in Education*, *24*(2), 227–243. doi:10.1108/QAE-01-2014-0004

Zadnik, K., Jones, L. A., Irvin, B. C., Kleinstein, R. N., Manny, R. E., Shin, J. A., & Mutti, D. O. (2000). Vision: Myopia and ambient night-time lighting. *Nature*, *404*(6774), 143–144. doi:10.1038/35004661 PMID:10724157

Zimbardo, P. G., & Boyd, J. N. (2015). Putting time in perspective: A valid, reliable individual- differences metric. *Journal of Personality and Social Psychology*, *77*(6), 1271–1288. doi:10.1037/0022-3514.77.6.1271

Chapter 7
Concluding Remarks

ABSTRACT

In this chapter, we provide brief concluding remarks which focus on summarizing key frameworks discussed throughout the first six chapters. We argue that by developing a testable model of statistics cognition we can make specific predictions about learning and errors which can help provide educators with the guidance needed to curb the influence errors can have on learning. Additionally, by developing a cognitive curriculum, which is designed to assess student affect and make use of meaningful individual differences, we can enhance the quality of statistics education in online environments. We conclude this chapter by outlining several future directions that can help facilitate our understanding of statistics cognition.

Technology has led to numerous changes in statistical practice... These changes in statistical practice have a direct impact on the content that should be taught, even in introductory material. - Chance, Ben-Zvi, Garfield, & Medina, 2007, p. 2

1. DEVELOPING A COGNITIVE ORIENTED STATISTICS COURSE

This text has taken the view that the purpose of statistics education is to enhance statistics cognition (Chapter 1). We have argued that to enhance statistics cognition we must have a clear model that allows us to make specific

DOI: 10.4018/978-1-5225-2420-5.ch007

predictions about the type of content statistics classes should offer and how they are designed, based on evidence supported methods to enhance cognitive development (Chapter 2). With a specific model of statistics cognition it is not only possible to understand how students learn statistics, but can also provide insight into the types of errors students make, allowing the learning process to become more efficient (Chapter 3). Taken together, the information afforded by a specific model of statistics cognition allows educators to develop a curriculum oriented on those cognitions (Chapter 4). However, in addition to teaching statistics content, educators must be aware of the role affect plays in learning, and develop procedures to test and correct harmful affective states (Chapter 5). To do this, educators and scholars must develop meaningful frameworks for exploring individual differences (Chapter 6).

Collectively, we propose an integrative framework for understanding the relationship of course features with learning processes when using a cognitive curriculum (Figure 1). In particular, this framework dictates that evidence is collected at each stage of the learning process that is used to inform specific hypotheses about student learning.

As depicted in Figure 1, any one unit has types of literacy, reasoning, and thinking objectives that are specific to material categorized within a unit. Before the course even begins, the educator needs to ensure that these cognitive learning objectives are aligned with appropriate assessments, and remain in the same relative proportion throughout each stage (i.e., if 30% of all learning objectives in a unit are based on literacy, then roughly 30% of all questions on the summative exam should measure literacy outcomes).

During the course the educator must monitor and evaluate student success. For instance, the educator must ensure students are following the correct procedures, such as mastering literacy before tackling thinking objectives. By giving learning a specific progression, students may begin to realize what to expect on exams (potentially reducing negative affective states) and better understand how they can succeed on assessments. For example, if students know that the majority of questions on an exam are based on recall, a strategy such as making flash cards could be effective. However, if most questions are on reasoning, then students may realize elaborative rehearsal works better (see Dunlosky, Rawson, Marsh, Nathan, & Willingham, 2013). Further, the educator must collect baseline information about student cognitive ability and affective responses to materials so that curriculum may be adjusted in real time. As outlined throughout this book, there are a variety of ways to improve student affect (Chapter 5) and potentially enhance motivation (Chapter 6). We argue that by collecting baseline information educators can

Concluding Remarks

Figure 1. Model of a cognitive statistics educational course in an online environment

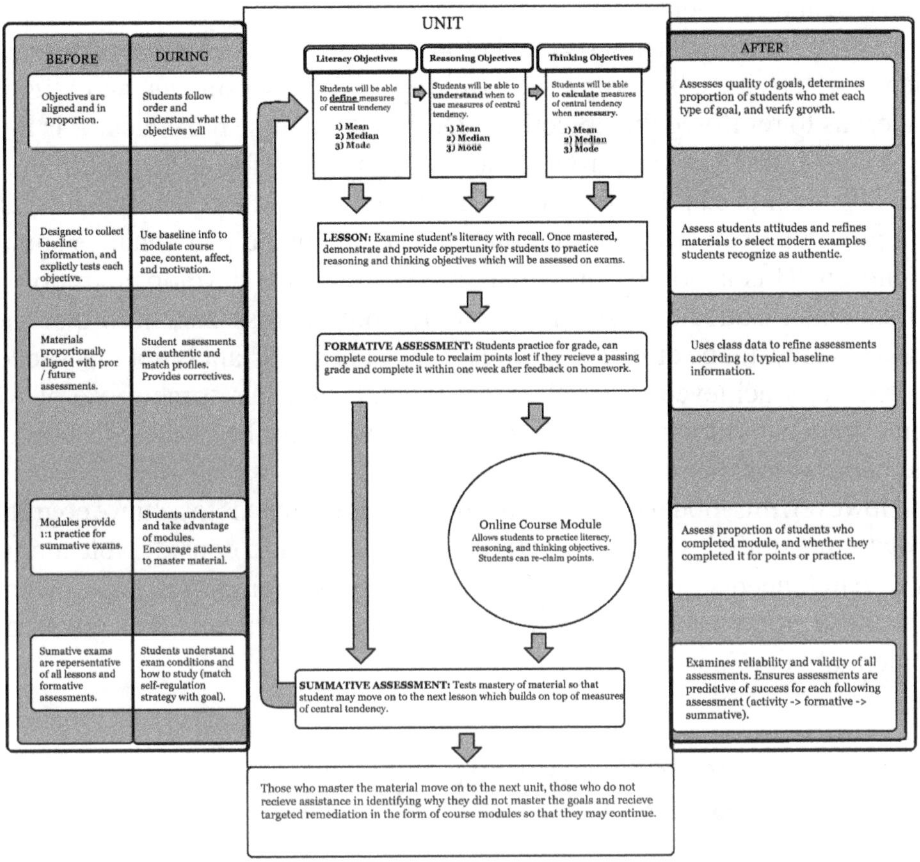

adjust the tone, pace, or even content depending upon the cognitive ability of students that term. Once the student has taken the summative exam for the unit, their progress can be assessed, and a decision about the student's ability to progress toward the next unit can be made.

At the end of the course, the educator can critically assess the efficacy of their curriculum. In addition to examining how many students can meet each goal the educator has a chance to verify their educational methods and look for cognitive development over the course of the assessments. This information, along with the baseline data collected about cognitive ability and attitudes, allows the educator to refine their materials for the next time they teach the course. Further, but assessing specific hypotheses, such as whether completion of course modules helped predict success, allows the educator to verify efficacy of curriculum. By assessing the psychometric properties of all activities, educators can refine their curriculum to be more effective and efficient.

Largely, this model asserts that before the course the educator is primarily planning activities, assessing face validity of all assessments, and aligning materials to reinforce the objectives. During the course the educator is there to provide motivation and troubleshooting assistance for student learning, so that all students can develop their cognitions. After the course the educator is a program evaluator, proving that their methods helped enhance student cognition. This is not to imply that an educator must develop cognition for all students. Educators will undoubtedly come across instances that their methods may not be effective. However, what is important is not that success is uniformly achieved, but rather that the educator has some sense of why some students did not achieve success, and what could be done to improve their course for the future.

However, this model is simply but one theoretical means to achieve cognitive growth in a statistics classroom. Importantly, there is a considerable amount of research needed to refine models for statistics cognition (Chapter 2) and how these processes could interact with curriculum structure (Chapter 4) and student affect (Chapter 5).

2. FUTURE DIRECTIONS FOR STATISTICS COGNITION RESEARCH

Statistics cognition is the basis for the new statistics education. Yet the definition of these constructs must be clarified in a manner accessible to both researchers and educators. While we attempted to provide such a model here, there are a variety of ways that the connection among the cognitive features of these constructs could be related (see Rumsey 2001; delMas, 2002). The first step in furthering the field of statistics education will be to begin assessing these models of statistics cognition, and developing ways that assessments can be shaped not just to measure cognitive growth, but to encourage it.

Additionally, while the study of online and statistics education have been studied for decades (Chapter 1), it is still worthwhile asking how statistics courses in online environments may be unique. While few studies have examined these differences (Summers, Waigandt, & Whittaker, 2005), there may be some reason to think that there could be unique implications for teaching statistics online. For instance, statistics courses are traditionally viewed in a negative light (Chapter 5; 3.2.1), that could be problematic for students who also have a low opinion of online learning (Summers et al., 2005).

Indeed, from a cognitive perspective, statistics may even be perceived as more difficult because students are often required to learn mathematical operations and formula on top of language and terms, suggesting that statistics students may have a higher cognitive burden. Indeed, a course that focuses on both phonological and visou-spatial material simultaneously may suggest that executive control is split among both the visuo-spatial sketch pad and phonological loop (Baddeley, 1986). Such a finding would be in line with prior research that has found that statistics anxiety is similar but distinct from math anxiety (Zeidner, 1991; Onwuegbuzie & Wilson, 2003), as well as work that scholars have interpreted to suggest that learning statistics is much like learning a language (Lalonde & Gardner, 1993; Saffran, 2003).

Further more research is needed to explore how statistics attitudes form, with particular emphasis on statistical mood as an underlying feature of these attitudes. Although considerable research has examined ways to describe and measure affective variables (Gal, Ginsburg, & Schau, 1997; Hood, Creed, Neumann, 2012; Schau, Stevens, Dauphinee, & Vecchio, 1995), there may be an overreliance in both statistics and online education research to focus on anxiety, and valence models of statistics emotions, as opposed to how affect can influence cognitions and decisions. In particular, future scholars must look at developing comprehensive affective models that informs the assessment and interpretation of statistics cognitions.

Finally, scholarly efforts should be directed toward identifying meaningful individual differences that inform the cognitive learning process (Chapter 6). We outlined how motivation can exist as a trait, and discussed implication for how individual trait motivation can be linked to macro-level organization features that can influence student perceptions of their environment. By identifying individual differences that describe, as opposed to label, students, we will begin to have a more meaningful understanding of how cognition may be different among students, and how education can be reformed to address these differences.

The field of statistics cognition largely is at its infancy as it relates to online technology. We hope that this book has provided a thoughtful synthesis of research that both useful and usable to both scholars and educators. We believe that focusing on statistical learning methods on developing statistics cognition will help educate and produce intelligent and responsible data scientists and data science students that will meet the demands of the modern data-driven society.

REFERENCES

Baddeley, A. D. (1986). *Working memory*. New York, NY: Oxford University Press.

delMas, R. C. (2002). Statistical literacy, reasoning, and learning: A commentary. *Journal of Statistics Education, 10*, 1–18. Retrieved from http://ww2.amstat.org/publications/jse/v10n3/delmas_discussion.html

Dunlosky, J., Rawson, K. A., Marsh, E. J., Nathan, M. J., & Willingham, D. T. (2013). Improving students learning with effective learning techniques: Promising directions from cognitive and educational psychology. *Psychological Science in the Public Interest, 14*(1), 4–58. doi:10.1177/1529100612453266 PMID:26173288

Gal, I., Ginsburg, L., & Schau, C. (1997). Monitoring attitudes and beliefs in statistics education. In *The assessment challenge in statistics education* (Vol. 12, pp. 37-51). Amsterdam, Netherlands: IOS Press.

Hood, M., Creed, P. A., & Neumann, D. L. (2012). Using the expectancy value model of motivation to understand the relationship between student attitudes and achievement in statistics. *Statistics Education Research Journal, 11*, 72–85.

Lalonde, R. N., & Gardner, R. C. (1993). Statistics as a second language? A model for predicting performance in psychology students. *Canadian Journal of Behavioural Science/Revue canadienne des sciences du comportement, 25*, 108-125.

Onwuegbuzie, A. J., & Wilson, V. A. (2003). Statistics Anxiety: Nature, etiology, antecedents, effects, and treatmentsa comprehensive review of the literature. *Teaching in Higher Education, 8*(2), 195–209. doi:10.1080/1356251032000052447

Rumsey, D. J. (2002). Statistical literacy as a goal for introductory statistics courses. *Journal of Statistics Education, 10*, 6–13. Retrieved from https://ww2.amstat.org/publications/jse/v10n3/rumsey2.html

Saffran, J. R. (2003). Statistical language learning mechanisms and constraints. *Current Directions in Psychological Science, 12*(4), 110–114. doi:10.1111/1467-8721.01243

Schau, C., Stevens, J., Dauphinee, T. L., & Vecchio, A. D. (1995). The development and validation of the survey of attitudes toward statistics. *Educational and Psychological Measurement*, *55*(5), 868–875. doi:10.1177/0013164495055005022

Summers, J. J., Waigandt, A., & Whittaker, T. A. (2005). A comparison of student achievement and satisfaction in an online versus a traditional face-to-face statistics class. *Innovative Higher Education*, *29*(3), 233–250. doi:10.1007/s10755-005-1938-x

Zeidner, M. (1991). Statistics and mathematics anxiety in social science students: Some interesting parallels. *The British Journal of Educational Psychology*, *61*(3), 319–328. doi:10.1111/j.2044-8279.1991.tb00989.x PMID:1786211

Related Readings

To continue IGI Global's long-standing tradition of advancing innovation through emerging research, please find below a compiled list of recommended IGI Global book chapters and journal articles in the areas of student cognition, online learning, and curriculum. These related readings will provide additional information and guidance to further enrich your knowledge and assist you with your own research.

Abdelaziz, H. A. (2014). Creative Design of Interactive eLearning Activities and Materials (IEAM): A Psycho-Pedagogical Model. *International Journal of Technology Diffusion*, *5*(4), 14–34. doi:10.4018/ijtd.2014100102

Abdelmalak, M. M., & Parra, J. L. (2016). Expanding Learning Opportunities for Graduate Students with HyFlex Course Design. *International Journal of Online Pedagogy and Course Design*, *6*(4), 19–37. doi:10.4018/IJOPCD.2016100102

Abera, B. (2014). Applying a Technological Pedagogical Content Knowledge Framework in Ethiopian English Language Teacher Education. In T. Issa, P. Isaias, & P. Kommers (Eds.), *Multicultural Awareness and Technology in Higher Education: Global Perspectives* (pp. 286–301). Hershey, PA: IGI Global. doi:10.4018/978-1-4666-5876-9.ch014

Adegbenro, J. B., Gumbo, M. T., & Olugbara, O. O. (2015). Exploring Technological Knowledge of Office Data Processing Teachers: Using Factor Analytic Methods. In M. Niess & H. Gillow-Wiles (Eds.), *Handbook of Research on Teacher Education in the Digital Age* (pp. 548–576). Hershey, PA: IGI Global. doi:10.4018/978-1-4666-8403-4.ch021

Amador, F., Nobre, A., & Barros, D. (2016). Towards a Model of a Didactics of eLearning: An Application to Education for Sustainable Development. In M. Pinheiro & D. Simões (Eds.), *Handbook of Research on Engaging Digital Natives in Higher Education Settings* (pp. 396–415). Hershey, PA: IGI Global. doi:10.4018/978-1-5225-0039-1.ch019

Amador, J. M., Kimmons, R., Miller, B. G., Desjardins, C. D., & Hall, C. (2015). Preparing Preservice Teachers to Become Self-Reflective of Their Technology Integration Practices. In M. Niess & H. Gillow-Wiles (Eds.), *Handbook of Research on Teacher Education in the Digital Age* (pp. 81–107). Hershey, PA: IGI Global. doi:10.4018/978-1-4666-8403-4.ch004

Archambault, L. (2014). Teaching Virtually: Strategies and Challenges in the 21st Century Online Classroom. *International Journal of Online Pedagogy and Course Design*, *4*(1), 1–15. doi:10.4018/ijopcd.2014010101

Arinze, B., Sylla, C., & Amobi, O. (2016). Cloud Computing for Teaching and Learning: Design Strategies. In L. Chao (Ed.), *Handbook of Research on Cloud-Based STEM Education for Improved Learning Outcomes* (pp. 159–171). Hershey, PA: IGI Global. doi:10.4018/978-1-4666-9924-3.ch011

Aşık, A. (2016). Digital Storytelling and Its Tools for Language Teaching: Perceptions and Reflections of Pre-Service Teachers. *International Journal of Computer-Assisted Language Learning and Teaching*, *6*(1), 55–68. doi:10.4018/IJCALLT.2016010104

Attard, C. (2015). Introducing iPads into Primary Mathematics Classrooms: Teachers' Experiences and Pedagogies. In M. Meletiou-Mavrotheris, K. Mavrou, & E. Paparistodemou (Eds.), *Integrating Touch-Enabled and Mobile Devices into Contemporary Mathematics Education* (pp. 193–213). Hershey, PA: IGI Global. doi:10.4018/978-1-4666-8714-1.ch009

Baert, H. (2015). Technology Tools, Proficiency, and Integration of Physical Education Teacher Educators. In M. Niess & H. Gillow-Wiles (Eds.), *Handbook of Research on Teacher Education in the Digital Age* (pp. 223–254). Hershey, PA: IGI Global. doi:10.4018/978-1-4666-8403-4.ch009

Banas, J. R., & York, C. S. (2014). The Impact of Authentic Learning Exercises On Pre-service Teachers Motivational Beliefs towards Technology Integration. *International Journal of Information and Communication Technology Education*, *10*(3), 60–76. doi:10.4018/ijicte.2014070105

Barakat, M., & Weiss-Randall, D. N. (2016). Through the Eyes of Students and Faculty: A Conceptual Framework for the Development of Online Courses. In V. Wang (Ed.), *Handbook of Research on Learning Outcomes and Opportunities in the Digital Age* (pp. 557–584). Hershey, PA: IGI Global. doi:10.4018/978-1-4666-9577-1.ch025

Baran, M. L., & Jones, J. E. (2014). Online Learning: Guidelines for Team Effectiveness. In C. Stevenson & J. Bauer (Eds.), *Building Online Communities in Higher Education Institutions: Creating Collaborative Experience* (pp. 279–292). Hershey, PA: IGI Global. doi:10.4018/978-1-4666-5178-4.ch015

Barbour, M. K., Adelstein, D., & Morrison, J. (2014). The Forgotten Teachers in K-12 Online Learning: Examining the Perceptions of Teachers Who Develop K-12 Online Courses. *International Journal of Online Pedagogy and Course Design*, *4*(3), 18–33. doi:10.4018/ijopcd.2014070102

Boboc, M. (2014). Connecting Communication to Curriculum and Pedagogy in Online Environments. In C. Stevenson & J. Bauer (Eds.), *Building Online Communities in Higher Education Institutions: Creating Collaborative Experience* (pp. 132–156). Hershey, PA: IGI Global. doi:10.4018/978-1-4666-5178-4.ch007

Borasi, R., Fredericksen, E., & Miller, D. (2017). From 0 to 60: The Case Study of a School of Education's Successful "Online Start-Up". In K. Shelton & K. Pedersen (Eds.), *Handbook of Research on Building* (pp. 60–83). Hershey, PA: IGI Global: Growing, and Sustaining Quality E-Learning Programs. doi:10.4018/978-1-5225-0877-9.ch004

Bowers, J., & Kumar, P. (2015). Students Perceptions of Teaching and Social Presence: A Comparative Analysis of Face-to-Face and Online Learning Environments. *International Journal of Web-Based Learning and Teaching Technologies*, *10*(1), 27–44. doi:10.4018/ijwltt.2015010103

Bozoglan, B., & Demirer, V. (2015). The Association between Internet Addiction and Psychosocial Variables. In J. Bishop (Ed.), *Psychological and Social Implications Surrounding Internet and Gaming Addiction* (pp. 171–185). Hershey, PA: IGI Global: IGI Global. doi:10.4018/978-1-4666-8595-6.ch010

Breen, P. (2014). Philosophies, Traditional Pedagogy, and New Technologies: A Report on a Case Study of EAP Teachers' Integration of Technology into Traditional Practice. In P. Breen (Ed.), *Cases on Teacher Identity* (pp. 317–341). Hershey, PA: IGI Global: Diversity, and Cognition in Higher Education. doi:10.4018/978-1-4666-5990-2.ch013

Brockmeier, F. (2017). Service Learning Online: Preparing to Work in Global Societies with E-Service-Learning. In C. Crosby & F. Brockmeier (Eds.), *Student Experiences and Educational Outcomes in Community Engagement for the 21st Century* (pp. 150–172). Hershey, PA: IGI Global. doi:10.4018/978-1-5225-0874-8.ch007

Brunvand, S. (2016). Facilitating Student Interaction and Collaboration in a MOOC Environment. In R. Mendoza-Gonzalez (Ed.), *User-Centered Design Strategies for Massive Open Online Courses (MOOCs)* (pp. 1–14). Hershey, PA: IGI Global. doi:10.4018/978-1-4666-9743-0.ch001

Bull, P. H., & Patterson, G. C. (2016). Strategies to Promote Pedagogical Knowledge Interplay with Technology. In J. Keengwe & G. Onchwari (Eds.), *Handbook of Research on Active Learning and the Flipped Classroom Model in the Digital Age* (pp. 255–271). Hershey, PA: IGI Global. doi:10.4018/978-1-4666-9680-8.ch013

Bush, S. B., Driskell, S. O., Niess, M. L., Pugalee, D., Rakes, C. R., & Ronau, R. N. (2015). The Impact of Digital Technologies in Mathematics Pre-Service Teacher Preparation over Four Decades. In M. Niess & H. Gillow-Wiles (Eds.), *Handbook of Research on Teacher Education in the Digital Age* (pp. 1–27). Hershey, PA: IGI Global. doi:10.4018/978-1-4666-8403-4.ch001

Byrd-Poller, L., Farmer, J. L., & Ford, V. (2017). The Role of Leaders in Facilitating Healing After Organizational Trauma. In S. Háša & R. Brunet-Thornton (Eds.), *Impact of Organizational Trauma on Workplace Behavior and Performance* (pp. 318–340). Hershey, PA: IGI Global: IGI Global. doi:10.4018/978-1-5225-2021-4.ch014

Cacho-Elizondo, S., Shahidi, N., & Tossan, V. (2015). Giving Up Smoking Using SMS Messages on your Mobile Phone. In A. Mesquita & C. Tsai (Eds.), *Human Behavior, Psychology, and Social Interaction in the Digital Era* (pp. 72–94). Hershey, PA: IGI Global: IGI Global. doi:10.4018/978-1-4666-8450-8.ch004

Card, S., & Wang, H. (2015). Taking Care to Play: Meaningful Communication in Dementia Care in Chinese Culture. In C. Dick-Muehlke, R. Li, & M. Orleans (Eds.), *Psychosocial Studies of the Individual's Changing Perspectives in Alzheimer's Disease* (pp. 76–103). Hershey, PA: IGI Global: IGI Global. doi:10.4018/978-1-4666-8478-2.ch004

Carrasco, G., & Kinnamon, E. (2017). An Examination of Selfish and Selfless Motives: A Review of the Social Psychological and Behavioral Economics Literature. In R. Ianole (Ed.), *Applied Behavioral Economics Research and Trends* (pp. 93–109). Hershey, PA: IGI Global: IGI Global. doi:10.4018/978-1-5225-1826-6.ch006

Cejka, P., & Mohelska, H. (2017). National Culture Influence on Organisational Trauma: A Conceptual Framework Review. In S. Hăša & R. Brunet-Thornton (Eds.), *Impact of Organizational Trauma on Workplace Behavior and Performance* (pp. 162–186). Hershey, PA: IGI Global: IGI Global. doi:10.4018/978-1-5225-2021-4.ch007

Chen, C. (2016). Effective Learning Strategies for the 21st Century: Implications for the E-Learning. In M. Anderson & C. Gavan (Eds.), *Developing Effective Educational Experiences through Learning Analytics* (pp. 143–169). Hershey, PA: IGI Global. doi:10.4018/978-1-4666-9983-0.ch006

Chen, C., Chiu, P., & Huang, Y. (2015). The Learning Style-Based Adaptive Learning System Architecture. *International Journal of Online Pedagogy and Course Design*, *5*(2), 1–10. doi:10.4018/IJOPCD.2015040101

Chen, R., Lin, T., & Xie, T. (2015). Towards Intelligent Window Layout Management: The Role of Mental Map. In A. Mesquita & C. Tsai (Eds.), *Human Behavior, Psychology, and Social Interaction in the Digital Era* (pp. 146–161). Hershey, PA: IGI Global: IGI Global. doi:10.4018/978-1-4666-8450-8.ch007

Cheng, H. (2017). A TL-TPACK Model on CSL Pre-Service Teachers' Competencies of Online Instruction. In C. Lin, D. Zhang, & B. Zheng (Eds.), *Preparing Foreign Language Teachers for Next-Generation Education* (pp. 198–225). Hershey, PA: IGI Global. doi:10.4018/978-1-5225-0483-2.ch011

Chuang, S., Lin, S., Chang, T., & Kaewmeesri, R. (2017). Behavioral Intention of Using Social Networking Site: A Comparative Study of Taiwanese and Thai Facebook Users. *International Journal of Technology and Human Interaction*, *13*(1), 61–81. doi:10.4018/IJTHI.2017010104

Cialdella, V. T., Lobato, E. J., & Jordan, J. S. (2017). Wild Architecture: Explaining Cognition via Self-Sustaining Systems. In J. Vallverdú, M. Mazzara, M. Talanov, S. Distefano, & R. Lowe (Eds.), *Advanced Research on Biologically Inspired Cognitive Architectures* (pp. 41–62). Hershey, PA: IGI Global. doi:10.4018/978-1-5225-1947-8.ch003

Çınar, M., & Tüzün, H. (2016). Web-Based Course Design Models. In M. Pinheiro & D. Simões (Eds.), *Handbook of Research on Engaging Digital Natives in Higher Education Settings* (pp. 374–395). Hershey, PA: IGI Global. doi:10.4018/978-1-5225-0039-1.ch018

Cleve, R. A., Işık, İ., & Pecanha, V. D. (2017). Sexual Identities in the Workplace: Avoiding Organizational Trauma When Disclosure Occurs – Current Perspectives. In S. Háša & R. Brunet-Thornton (Eds.), *Impact of Organizational Trauma on Workplace Behavior and Performance* (pp. 188–220). Hershey, PA: IGI Global: IGI Global. doi:10.4018/978-1-5225-2021-4.ch008

Conrad, S. S., & Dabbagh, N. (2015). Examining the Factors that Influence how Instructors Provide Feedback in Online Learning Environments. *International Journal of Online Pedagogy and Course Design*, *5*(4), 47–66. doi:10.4018/IJOPCD.2015100104

Contier, A. T., & Torres, L. (2017). Neuroaesthetics: Insights into the Aesthetic Experience of Visual Art. In R. Zuanon (Ed.), *Projective Processes and Neuroscience in Art and Design* (pp. 87–102). Hershey, PA: IGI Global. doi:10.4018/978-1-5225-0510-5.ch006

Courey, S., LePage, P., Blackorby, J., Siker, J., & Nguyen, T. (2015). The Effects of Using Dynabook to Prepare Special Education Teachers to Teach Proportional Reasoning. *International Journal of Web-Based Learning and Teaching Technologies*, *10*(1), 45–64. doi:10.4018/ijwltt.2015010104

Cranton, P., & Thompson, P. (2014). Creating Collaboration in Global Online Learning: Case Studies. In V. Wang (Ed.), *Handbook of Research on Education and Technology in a Changing Society* (pp. 92–103). Hershey, PA: IGI Global. doi:10.4018/978-1-4666-6046-5.ch007

Crawford, C. M., & Smith, M. S. (2015). Rethinking Bloom's Taxonomy: Implicit Cognitive Vulnerability as an Impetus towards Higher Order Thinking Skills. In Z. Jin (Ed.), *Exploring Implicit Cognition: Learning, Memory, and Social Cognitive Processes* (pp. 86–103). Hershey, PA: IGI Global. doi:10.4018/978-1-4666-6599-6.ch004

de Groot, C., Fogleman, J., & Kern, D. (2015). Using Mobile Technologies to Co-Construct TPACK in Teacher Education. In J. Keengwe & M. Maxfield (Eds.), *Advancing Higher Education with Mobile Learning Technologies: Cases* (pp. 195–219). Hershey, PA: IGI Global. doi:10.4018/978-1-4666-6284-1.ch011

de Soir, E., & Kleber, R. (2017). Understanding the Core of Psychological Trauma: Trauma in Contemporary French Theory. In S. Háša & R. Brunet-Thornton (Eds.), *Impact of Organizational Trauma on Workplace Behavior and Performance* (pp. 57–75). Hershey, PA: IGI Global. doi:10.4018/978-1-5225-2021-4.ch003

Delgado, J. J. (2017). How Is the Personality of Facebook Customers?: Cloninger's Psychobiological Model of Temperament as a Predictor of SNSs. In M. Dos Santos (Ed.), *Applying Neuroscience to Business Practice* (pp. 191–229). Hershey, PA: IGI Global. doi:10.4018/978-1-5225-1028-4.ch009

Dentale, F., Vecchione, M., & Barbaranelli, C. (2015). Applying the IAT to Assess Big Five Personality Traits: A Brief Review of Measurement and Validity Issues. In Z. Jin (Ed.), *Exploring Implicit Cognition: Learning, Memory, and Social Cognitive Processes* (pp. 1–15). Hershey, PA: IGI Global. doi:10.4018/978-1-4666-6599-6.ch001

Deyoe, M. M., Newman, D. L., & Asaro-Saddler, K. (2014). Moving from Professional Development to Real-Time Use: How are we Changing Students? In J. Keengwe, G. Onchwari, & D. Hucks (Eds.), *Literacy Enrichment and Technology Integration in Pre-Service Teacher Education* (pp. 160–182). Hershey, PA: IGI Global. doi:10.4018/978-1-4666-4924-8.ch010

Dick, T. P., & Burrill, G. F. (2016). Design and Implementation Principles for Dynamic Interactive Mathematics Technologies. In M. Niess, S. Driskell, & K. Hollebrands (Eds.), *Handbook of Research on Transforming Mathematics Teacher Education in the Digital Age* (pp. 23–51). Hershey, PA: IGI Global. doi:10.4018/978-1-5225-0120-6.ch002

Díez, J. C., & Saiz-Alvarez, J. M. (2016). Leadership in Social Entrepreneurship: Is It Ability or Skill? In J. Saiz-Álvarez (Ed.), *Handbook of Research on Social Entrepreneurship and Solidarity Economics* (pp. 134–153). Hershey, PA: IGI Global. doi:10.4018/978-1-5225-0097-1.ch008

Dikici, A. (2017). Revisiting the Relationships between Turkish Prospective Teachers' Thinking Styles and Behaviors Fostering Creativity. In N. Silton (Ed.), *Exploring the Benefits of Creativity in Education, Media, and the Arts* (pp. 136–157). Hershey, PA: IGI Global. doi:10.4018/978-1-5225-0504-4.ch007

Dodd, B. J., Baukal, C. E. Jr, & Ausburn, L. J. (2016). A Post-Positivist Framework for Using and Building Theory in Online Instructional Design. *International Journal of Online Pedagogy and Course Design*, *6*(4), 53–70. doi:10.4018/IJOPCD.2016100104

Driskell, S. O., Bush, S. B., Ronau, R. N., Niess, M. L., Rakes, C. R., & Pugalee, D. K. (2016). Mathematics Education Technology Professional Development: Changes over Several Decades. In M. Niess, S. Driskell, & K. Hollebrands (Eds.), *Handbook of Research on Transforming Mathematics Teacher Education in the Digital Age* (pp. 107–136). Hershey, PA: IGI Global. doi:10.4018/978-1-5225-0120-6.ch005

Dunston, Y. L., Patterson, G. C., & Bull, P. H. (2016). Faculty Perspectives of Technology-Enhanced Course Redesign. In J. Keengwe & G. Onchwari (Eds.), *Handbook of Research on Active Learning and the Flipped Classroom Model in the Digital Age* (pp. 150–176). Hershey, PA: IGI Global. doi:10.4018/978-1-4666-9680-8.ch008

Duvall, M., Matranga, A., Foster, A., & Silverman, J. (2016). Mobile Learning: Technology as Mediator of Personal and School Experiences. *International Journal of Game-Based Learning*, *6*(1), 30–42. doi:10.4018/IJGBL.2016010103

Eapen, V., & Walter, A. (2016). Mind the Gap: Developmental Vulnerability and Mental Health. In R. Gopalan (Ed.), *Handbook of Research on Diagnosing, Treating, and Managing Intellectual Disabilities* (pp. 11–32). Hershey, PA: IGI Global. doi:10.4018/978-1-5225-0089-6.ch002

Edson, A. J., & Thomas, A. (2016). Transforming Preservice Mathematics Teacher Knowledge for and with the Enacted Curriculum: The Case of Digital Instructional Materials. In M. Niess, S. Driskell, & K. Hollebrands (Eds.), *Handbook of Research on Transforming Mathematics Teacher Education in the Digital Age* (pp. 215–240). Hershey, PA: IGI Global. doi:10.4018/978-1-5225-0120-6.ch009

Fallon, F. (2017). Integrated Information Theory (IIT) and Artificial Consciousness. In J. Vallverdú, M. Mazzara, M. Talanov, S. Distefano, & R. Lowe (Eds.), *Advanced Research on Biologically Inspired Cognitive Architectures* (pp. 1–23). Hershey, PA: IGI Global. doi:10.4018/978-1-5225-1947-8.ch001

Fang, L., & Ha, L. (2015). Do College Students Benefit from Their Social Media Experience?: Social Media Involvement and Its Impact on College Students' Self-Efficacy Perception. In A. Mesquita & C. Tsai (Eds.), *Human Behavior, Psychology, and Social Interaction in the Digital Era* (pp. 259–278). Hershey, PA: IGI Global. doi:10.4018/978-1-4666-8450-8.ch013

Farmer, L. (2014). Technology Impact on New Adult Behavior about Health Information. In V. Wang (Ed.), *Handbook of Research on Adult and Community Health Education: Tools, Trends, and Methodologies* (pp. 68–81). Hershey, PA: IGI Global. doi:10.4018/978-1-4666-6260-5.ch005

Fasko, D. (2017). Creativity in the Schools: Educational Changes Lately? In N. Silton (Ed.), *Exploring the Benefits of Creativity in Education, Media, and the Arts* (pp. 92–116). Hershey, PA: IGI Global. doi:10.4018/978-1-5225-0504-4.ch005

Fasso, W., Knight, C., & Knight, B. A. (2014). A Learner-Centered Design Framework for E-Learning. *International Journal of Online Pedagogy and Course Design, 4*(4), 44–59. doi:10.4018/ijopcd.2014100104

Fazio, S., & Mitchell, D. B. (2015). Self-Preservation in Individuals with Alzheimer's Disease: Empirical Evidence and the Role of the Social Environment. In C. Dick-Muehlke, R. Li, & M. Orleans (Eds.), *Psychosocial Studies of the Individual's Changing Perspectives in Alzheimer's Disease* (pp. 183–207). Hershey, PA: IGI Global. doi:10.4018/978-1-4666-8478-2.ch008

Feitosa-Santana, C. (2017). Understanding How the Mind Works: The Neuroscience of Perception, Behavior, and Creativity. In R. Zuanon (Ed.), *Projective Processes and Neuroscience in Art and Design* (pp. 239–252). Hershey, PA: IGI Global. doi:10.4018/978-1-5225-0510-5.ch014

Felger, J., & Shafer, K. G. (2016). An Algebra Teacher's Instructional Decision-Making Process with GeoGebra: Thinking with a TPACK Mindset. In M. Niess, S. Driskell, & K. Hollebrands (Eds.), *Handbook of Research on Transforming Mathematics Teacher Education in the Digital Age* (pp. 493–518). Hershey, PA: IGI Global. doi:10.4018/978-1-5225-0120-6.ch019

Ferris, A. (2017). Creativity in the Emerging Adult. In N. Silton (Ed.), *Exploring the Benefits of Creativity in Education, Media, and the Arts* (pp. 26–49). Hershey, PA: IGI Global. doi:10.4018/978-1-5225-0504-4.ch002

Fife, E., Nelson, C. L., & Clarke, T. B. (2014). Online Technological Media in the Higher Education Classroom: An Exploratory Investigation of Varied Levels of Twitter Use. *International Journal of Online Pedagogy and Course Design*, *4*(2), 35–45. doi:10.4018/ijopcd.2014040103

Figg, C., & Jaipal-Jamani, K. (2015). Developing Teacher Knowledge about Gamification as an Instructional Strategy. In M. Niess & H. Gillow-Wiles (Eds.), *Handbook of Research on Teacher Education in the Digital Age* (pp. 663–690). Hershey, PA: IGI Global. doi:10.4018/978-1-4666-8403-4.ch025

Finger, G. (2015). Improving Initial Teacher Education in Australia: Solutions and Recommendations from the Teaching Teachers for the Future Project. In N. Ololube, P. Kpolovie, & L. Makewa (Eds.), *Handbook of Research on Enhancing Teacher Education with Advanced Instructional Technologies* (pp. 187–207). Hershey, PA: IGI Global. doi:10.4018/978-1-4666-8162-0.ch010

Flores, A., Park, J., & Bernhardt, S. A. (2016). Learning Mathematics and Technology through Inquiry, Cooperation, and Communication: A Learning Trajectory for Future Mathematics Teachers. In M. Niess, S. Driskell, & K. Hollebrands (Eds.), *Handbook of Research on Transforming Mathematics Teacher Education in the Digital Age* (pp. 324–352). Hershey, PA: IGI Global. doi:10.4018/978-1-5225-0120-6.ch013

Fogliano, F., & Oliveira, H. C. (2017). Neuroesthetics: Perspectives and Reflections. In R. Zuanon (Ed.), *Projective Processes and Neuroscience in Art and Design* (pp. 52–70). Hershey, PA: IGI Global. doi:10.4018/978-1-5225-0510-5.ch004

Fokides, E. (2016). Pre-Service Teachers, Computers, and ICT Courses: A Troubled Relationship. *International Journal of Information and Communication Technology Education, 12*(4), 25–36. doi:10.4018/IJICTE.2016100103

Folk, J. R., & Eskenazi, M. A. (2017). Eye Movement Behavior and Individual Differences in Word Identification During Reading. In C. Was, F. Sansosti, & B. Morris (Eds.), *Eye-Tracking Technology Applications in Educational Research* (pp. 66–87). Hershey, PA: IGI Global. doi:10.4018/978-1-5225-1005-5.ch004

Foster, A., & Shah, M. (2016). Knew Me and New Me: Facilitating Student Identity Exploration and Learning through Game Integration. *International Journal of Gaming and Computer-Mediated Simulations, 8*(3), 39–58. doi:10.4018/IJGCMS.2016070103

Franco, M., Ortiz, T. V., Amorim, H. A., & Faber, J. (2017). Can We Induce a Cognitive Representation of a Prosthetic Arm by Means of Crossmodal Stimuli? In R. Zuanon (Ed.), *Projective Processes and Neuroscience in Art and Design* (pp. 182–204). Hershey, PA: IGI Global. doi:10.4018/978-1-5225-0510-5.ch011

Frazier, L. C., & Palmer, B. M. (2015). Effective Online Learning Begins with Effective Teacher Preparation. In T. Heafner, R. Hartshorne, & T. Petty (Eds.), *Exploring the Effectiveness of Online Education in K-12 Environments* (pp. 148–168). Hershey, PA: IGI Global. doi:10.4018/978-1-4666-6383-1.ch008

Gallagher-Lepak, S., & Vandenhouten, C. (2016). E-Learning and Faculty Development in Higher Education: A Comprehensive Project. In B. Khan (Ed.), *Revolutionizing Modern Education through Meaningful E-Learning Implementation* (pp. 226–244). Hershey, PA: IGI Global. doi:10.4018/978-1-5225-0466-5.ch012

Gallego, J. (2017). Organizational Trauma and Change Management. In S. Háša & R. Brunet-Thornton (Eds.), *Impact of Organizational Trauma on Workplace Behavior and Performance* (pp. 140–161). Hershey, PA: IGI Global. doi:10.4018/978-1-5225-2021-4.ch006

Galvin, J. E., & Kelleher, M. E. (2015). Dementia and Other Neurocognitive Disorders: An Overview. In C. Dick-Muehlke, R. Li, & M. Orleans (Eds.), *Psychosocial Studies of the Individual's Changing Perspectives in Alzheimer's Disease* (pp. 104–130). Hershey, PA: IGI Global. doi:10.4018/978-1-4666-8478-2.ch005

Gardner, M. K., & Strayer, D. L. (2017). What Cognitive Psychology Can Tell Us About Educational Computer Games. In R. Zheng & M. Gardner (Eds.), *Handbook of Research on Serious Games for Educational Applications* (pp. 1–18). Hershey, PA: IGI Global. doi:10.4018/978-1-5225-0513-6.ch001

Garg, B., Khanna, P., & Khanna, A. (2017). Chronic Mental Illness and the Changing Scope of Intervention Strategies, Diagnosis, and Treatment in Child and Adolescent Population. In B. Prasad (Ed.), *Chronic Mental Illness and the Changing Scope of Intervention Strategies, Diagnosis, and Treatment* (pp. 258–269). Hershey, PA: IGI Global. doi:10.4018/978-1-5225-0519-8.ch014

Gautreau, C. S., Stang, K. K., Street, C., & Guillaume, A. (2014). Making the Move: Supporting Faculty in the Transition to Blended or Online Courses. *International Journal of Online Pedagogy and Course Design*, *4*(1), 27–42. doi:10.4018/ijopcd.2014010103

Giannouli, V. (2017). Creativity and Giftedness: A Study of Attitudes. In N. Silton (Ed.), *Exploring the Benefits of Creativity in Education, Media, and the Arts* (pp. 179–197). Hershey, PA: IGI Global. doi:10.4018/978-1-5225-0504-4.ch009

Gikandi, J. W. (2015). Towards a Theory of Formative Assessment in Online Higher Education. In J. Keengwe (Ed.), *Handbook of Research on Educational Technology Integration and Active Learning* (pp. 292–316). Hershey, PA: IGI Global. doi:10.4018/978-1-4666-8363-1.ch014

Gikandi, J. W. (2017). Computer-Supported Collaborative Learning and Assessment: A Strategy for Developing Online Learning Communities in Continuing Education. In J. Keengwe & G. Onchwari (Eds.), *Handbook of Research on Learner-Centered Pedagogy in Teacher Education and Professional Development* (pp. 309–333). Hershey, PA: IGI Global. doi:10.4018/978-1-5225-0892-2.ch017

Gillow-Wiles, H., & Niess, M. L. (2015). Engaging Google Docs to Support Collaboration and Reflection in Online Teacher Education. In M. Niess & H. Gillow-Wiles (Eds.), *Handbook of Research on Teacher Education in the Digital Age* (pp. 635–662). Hershey, PA: IGI Global. doi:10.4018/978-1-4666-8403-4.ch024

Gillow-Wiles, H., & Niess, M. L. (2016). A Reconstructed Conception of Learner Engagement in Technology Rich Online Learning Environments. In M. Niess, S. Driskell, & K. Hollebrands (Eds.), *Handbook of Research on Transforming Mathematics Teacher Education in the Digital Age* (pp. 577–607). Hershey, PA: IGI Global. doi:10.4018/978-1-5225-0120-6.ch022

Golden, J. E., & Brown, V. (2017). A Holistic Professional Development Model: A Case Study to Support Faculty Transition to Online Teaching. In C. Martin & D. Polly (Eds.), *Handbook of Research on Teacher Education and Professional Development* (pp. 259–284). Hershey, PA: IGI Global. doi:10.4018/978-1-5225-1067-3.ch014

Goodson-Espy, T., & Poling, L. (2015). Interactive Whiteboards: Preparing Secondary Mathematics Teachers to Avoid Catch-22. In D. Polly (Ed.), *Cases on Technology Integration in Mathematics Education* (pp. 288–307). Hershey, PA: IGI Global. doi:10.4018/978-1-4666-6497-5.ch014

Gopalan, R. T. (2016). Intellectual Disability: From History to Recent Trends. In R. Gopalan (Ed.), *Handbook of Research on Diagnosing, Treating, and Managing Intellectual Disabilities* (pp. 1–10). Hershey, PA: IGI Global. doi:10.4018/978-1-5225-0089-6.ch001

Guan, S. (2014). Internet-Based Technology Use in Second Language Learning: A Systematic Review. *International Journal of Cyber Behavior, Psychology and Learning*, *4*(4), 69–81. doi:10.4018/ijcbpl.2014100106

Gupta, S., Taneja, S., & Kumar, N. (2015). Redefining the Classroom: Integration of Open and Classroom Learning in Higher Education. In E. McKay & J. Lenarcic (Eds.), *Macro-Level Learning through Massive Open Online Courses (MOOCs)* (pp. 168–182). Hershey, PA: IGI Global. doi:10.4018/978-1-4666-8324-2.ch010

Hacker, D. J. (2017). The Role of Metacognition in Learning via Serious Games. In R. Zheng & M. Gardner (Eds.), *Handbook of Research on Serious Games for Educational Applications* (pp. 19–40). Hershey, PA: IGI Global. doi:10.4018/978-1-5225-0513-6.ch002

Hadjileontiadou, S. J., Dias, S. B., Diniz, J. A., & Hadjileontiadis, L. J. (2015). Understanding Online Learning Environments (OLEs). In Fuzzy Logic-Based Modeling in Collaborative and Blended Learning (pp. 18-50). Hershey, PA: IGI Global. doi:10.4018/978-1-4666-8705-9.ch002

Halder, S., & Mahato, A. (2017). Cognitive Remediation Therapy in Chronic Schizophrenia. In B. Prasad (Ed.), *Chronic Mental Illness and the Changing Scope of Intervention Strategies, Diagnosis, and Treatment* (pp. 292–307). Hershey, PA: IGI Global. doi:10.4018/978-1-5225-0519-8.ch016

Harkness, S. J. (2014). Program Administration and Implementation of an Online Learning Initiative at a Historically Black College University. In M. Orleans (Ed.), *Cases on Critical and Qualitative Perspectives in Online Higher Education* (pp. 44–60). Hershey, PA: IGI Global. doi:10.4018/978-1-4666-5051-0.ch003

Harrington, R. A., Driskell, S. O., Johnston, C. J., Browning, C. A., & Niess, M. L. (2016). Technological Pedagogical Content Knowledge: Preparation and Support of Mathematics Teachers. In M. Niess, S. Driskell, & K. Hollebrands (Eds.), *Handbook of Research on Transforming Mathematics Teacher Education in the Digital Age* (pp. 1–22). Hershey, PA: IGI Global. doi:10.4018/978-1-5225-0120-6.ch001

Heins, S., Heins, G., & Dick-Muehlke, C. (2015). Steve's Story: Living with Mild Cognitive Impairment. In C. Dick-Muehlke, R. Li, & M. Orleans (Eds.), *Psychosocial Studies of the Individual's Changing Perspectives in Alzheimer's Disease* (pp. 33–60). Hershey, PA: IGI Global. doi:10.4018/978-1-4666-8478-2.ch002

Hennessey, S., Olofson, M. W., Swallow, M. J., & Downes, J. M. (2015). Evolving Pedagogy and Practice: The 1:1 Mathematics Classroom through a TPACK Lens. In M. Niess & H. Gillow-Wiles (Eds.), *Handbook of Research on Teacher Education in the Digital Age* (pp. 577–603). Hershey, PA: IGI Global. doi:10.4018/978-1-4666-8403-4.ch022

Hilbert, E., & Mierau, K. (2014). Efficiency and Quality Improvement in Online Course Development. In M. Orleans (Ed.), *Cases on Critical and Qualitative Perspectives in Online Higher Education* (pp. 435–451). Hershey, PA: IGI Global. doi:10.4018/978-1-4666-5051-0.ch022

Hsien, O. L., Eak, A. D., Vighnarajah, S., Huah, G. L., & Teik, O. C. (2016). Qualitative Findings on the Dynamics of Online Facilitation in Distance Education. *International Journal of Online Pedagogy and Course Design*, *6*(4), 1–18. doi:10.4018/IJOPCD.2016100101

Huang, C., & Liu, E. Z. (2015). E-Tutor Perceptions towards the Star Rural Area E-Learning Project. *International Journal of Online Pedagogy and Course Design*, *5*(1), 20–29. doi:10.4018/ijopcd.2015010102

Hunter, J. L. (2015). High Possibility Classrooms: A New Model for Technology Integration. In M. Niess & H. Gillow-Wiles (Eds.), *Handbook of Research on Teacher Education in the Digital Age* (pp. 466–492). Hershey, PA: IGI Global. doi:10.4018/978-1-4666-8403-4.ch018

Hurtado, I. G., & Llamas, J. M. (2014). Social Networks in University Classrooms: An Experience of Teaching and Learning with Pre-Service Teachers through Facebook. *International Journal of Online Pedagogy and Course Design*, *4*(3), 34–48. doi:10.4018/ijopcd.2014070103

Isik, I. (2017). Organizations and Exposure to Trauma at a Collective Level: The Taxonomy of Potentially Traumatic Events. In S. Háša & R. Brunet-Thornton (Eds.), *Impact of Organizational Trauma on Workplace Behavior and Performance* (pp. 18–56). Hershey, PA: IGI Global. doi:10.4018/978-1-5225-2021-4.ch002

Jančec, L., Vorkapić, S. T., & Vodopivec, J. L. (2015). Hidden Curriculum Determinants in (Pre)School Institutions: Implicit Cognition in Action. In Z. Jin (Ed.), *Exploring Implicit Cognition: Learning, Memory, and Social Cognitive Processes* (pp. 216–242). Hershey, PA: IGI Global. doi:10.4018/978-1-4666-6599-6.ch011

Jang, J. E., & Lei, J. (2015). The Impact of Video Self-Analysis on the Development of Preservice Teachers Technological Pedagogical Content Knowledge (TPACK). *International Journal of Digital Literacy and Digital Competence*, *6*(4), 13–29. doi:10.4018/IJDLDC.2015100102

Jang, S., & Tsai, M. (2016). Exploring the Development of Pre-Service Teachers' ICT-TPACK using a Cognitive Stimulation Tool. In E. Railean, G. Walker, A. Elçi, & L. Jackson (Eds.), *Handbook of Research on Applied Learning Theory and Design in Modern Education* (pp. 380–404). Hershey, PA: IGI Global. doi:10.4018/978-1-4666-9634-1.ch018

Jerabek, I., & Muoio, D. (2017). The Stress Profile: The Influence of Personal Characteristics on Response to Occupational Trauma. In S. Háša & R. Brunet-Thornton (Eds.), *Impact of Organizational Trauma on Workplace Behavior and Performance* (pp. 77–119). Hershey, PA: IGI Global. doi:10.4018/978-1-5225-2021-4.ch004

Jha, S., Khanna, A., & Khanna, P. (2017). Advanced Intervention Strategies for Suicide in Patients with Chronic Mental Illness. In B. Prasad (Ed.), *Chronic Mental Illness and the Changing Scope of Intervention Strategies, Diagnosis, and Treatment* (pp. 271–291). Hershey, PA: IGI Global. doi:10.4018/978-1-5225-0519-8.ch015

Johard, L., Lippi, V., Safina, L., & Mazzara, M. (2017). Mind and Matter: Why It All Makes Sense. In J. Vallverdú, M. Mazzara, M. Talanov, S. Distefano, & R. Lowe (Eds.), *Advanced Research on Biologically Inspired Cognitive Architectures* (pp. 63–82). Hershey, PA: IGI Global. doi:10.4018/978-1-5225-1947-8.ch004

Johnson, G. M., & Cooke, A. (2016). An Ecological Model of Student Interaction in Online Learning Environments. In L. Kyei-Blankson, J. Blankson, E. Ntuli, & C. Agyeman (Eds.), *Handbook of Research on Strategic Management of Interaction* (pp. 1–28). Hershey, PA: IGI Global. doi:10.4018/978-1-4666-9582-5.ch001

Kasemsap, K. (2017). Investigating the Roles of Neuroscience and Knowledge Management in Higher Education. In S. Mukerji & P. Tripathi (Eds.), *Handbook of Research on Administration, Policy, and Leadership in Higher Education* (pp. 112–140). Hershey, PA: IGI Global. doi:10.4018/978-1-5225-0672-0.ch006

Kasemsap, K. (2017). Mastering Cognitive Neuroscience and Social Neuroscience Perspectives in the Information Age. In M. Dos Santos (Ed.), *Applying Neuroscience to Business Practice* (pp. 82–113). Hershey, PA: IGI Global. doi:10.4018/978-1-5225-1028-4.ch005

Kaushik, P., & Singh, T. B. (2017). Predictors of Expressed Emotion, Caregiver's Burden, and Quality of Life in Chronic Mental Illness. In B. Prasad (Ed.), *Chronic Mental Illness and the Changing Scope of Intervention Strategies, Diagnosis, and Treatment* (pp. 143–163). Hershey, PA: IGI Global. doi:10.4018/978-1-5225-0519-8.ch008

Keyser, A. K., & Corning, M. (2017). Creative Aging: Stimulating Creativity in Middle and Late Adulthood. In N. Silton (Ed.), *Exploring the Benefits of Creativity in Education, Media, and the Arts* (pp. 50–66). Hershey, PA: IGI Global. doi:10.4018/978-1-5225-0504-4.ch003

Khan, K. P. (2017). Improving the Quality of Online Learning Environments: The Value of an Online Specific Design Model. In K. Shelton & K. Pedersen (Eds.), *Handbook of Research on Building* (pp. 278–300). Hershey, PA: IGI Global. doi:10.4018/978-1-5225-0877-9.ch014

Kılıç, B. (2017). An Organizational Trauma Intervention: A Case From Turkey. In S. Háša & R. Brunet-Thornton (Eds.), *Impact of Organizational Trauma on Workplace Behavior and Performance* (pp. 264–277). Hershey, PA: IGI Global. doi:10.4018/978-1-5225-2021-4.ch011

Kirsch, C., Lubart, T., de Vries, H., & Houssemand, C. (2017). Scientific Creativity in Psychology: A Cognitive-Conative Approach. In C. Zhou (Ed.), *Handbook of Research on Creative Problem-Solving Skill Development in Higher Education* (pp. 51–73). Hershey, PA: IGI Global. doi:10.4018/978-1-5225-0643-0.ch003

Kleinmintz, O. M. (2017). Train Yourself to Let Go: The Benefits of Deliberate Practice on Creativity and Its Neural Basis. In N. Silton (Ed.), *Exploring the Benefits of Creativity in Education, Media, and the Arts* (pp. 67–90). Hershey, PA: IGI Global. doi:10.4018/978-1-5225-0504-4.ch004

Knutas, A., Ikonen, J., Maggiorini, D., Ripamonti, L., & Porras, J. (2016). Creating Student Interaction Profiles for Adaptive Collaboration Gamification Design. *International Journal of Human Capital and Information Technology Professionals*, *7*(3), 47–62. doi:10.4018/IJHCITP.2016070104

Kučera, D. (2017). The Potential of Spirituality for the Treatment of Organizational Trauma. In S. Háša & R. Brunet-Thornton (Eds.), *Impact of Organizational Trauma on Workplace Behavior and Performance* (pp. 295–317). Hershey, PA: IGI Global. doi:10.4018/978-1-5225-2021-4.ch013

Kukreti, P., Khanna, P., & Khanna, A. (2017). Chronic Mental Illnesses and Homelessness. In B. Prasad (Ed.), *Chronic Mental Illness and the Changing Scope of Intervention Strategies, Diagnosis, and Treatment* (pp. 1–20). Hershey, PA: IGI Global. doi:10.4018/978-1-5225-0519-8.ch001

Kumi-Yeboah, A. (2015). Learning Theory and Online Learning in K-12 Education: Instructional Models and Implications. In T. Heafner, R. Hartshorne, & T. Petty (Eds.), *Exploring the Effectiveness of Online Education in K-12 Environments* (pp. 126–146). Hershey, PA: IGI Global. doi:10.4018/978-1-4666-6383-1.ch007

Kuss, D. J. (2015). "I Can't Do It by Myself": An IPA of Clients Seeking Psychotherapy for Their MMORPG Addiction. In J. Bishop (Ed.), *Psychological and Social Implications Surrounding Internet and Gaming Addiction* (pp. 78–110). Hershey, PA: IGI Global. doi:10.4018/978-1-4666-8595-6.ch006

Landim, P. D. (2017). Design and Emotion: Contributions to the Emotional Design. In R. Zuanon (Ed.), *Projective Processes and Neuroscience in Art and Design* (pp. 119–136). Hershey, PA: IGI Global. doi:10.4018/978-1-5225-0510-5.ch008

Lebraty, J., & Godé, C. (2015). Assessing the Performance of Decision Support Systems in Military Environment: The 3C Method. In A. Mesquita & C. Tsai (Eds.), *Human Behavior, Psychology, and Social Interaction in the Digital Era* (pp. 45–70). Hershey, PA: IGI Global. doi:10.4018/978-1-4666-8450-8.ch003

Lee, K., & Brett, C. (2015). An Online Course Design for Inservice Teacher Professional Development in a Digital Age: The Effectiveness of the Double-Layered CoP Model. In M. Niess & H. Gillow-Wiles (Eds.), *Handbook of Research on Teacher Education in the Digital Age* (pp. 55–80). Hershey, PA: IGI Global. doi:10.4018/978-1-4666-8403-4.ch003

Lehenbauer-Baum, M., & Fohringer, M. (2015). Internet Gaming Disorder: A Deeper Look into Addiction vs. High Engagement. In J. Bishop (Ed.), *Psychological and Social Implications Surrounding Internet and Gaming Addiction* (pp. 1–15). Hershey, PA: IGI Global. doi:10.4018/978-1-4666-8595-6.ch001

Lei, M., Liu, W., Gao, Y., & Zhu, T. (2015). Mobile User Behaviors in China. In Z. Yan (Ed.), *Encyclopedia of Mobile Phone Behavior* (pp. 1110–1128). Hershey, PA: IGI Global. doi:10.4018/978-1-4666-8239-9.ch091

Leote, R. (2017). Perceptual Processes and Multisensoriality: Understanding Multimodal Art from Neuroscientific Concepts. In R. Zuanon (Ed.), *Projective Processes and Neuroscience in Art and Design* (pp. 1–14). Hershey, PA: IGI Global. doi:10.4018/978-1-5225-0510-5.ch001

Limone, P., & Pace, R.Revolution in the School. (2016). Teacher Training and Digital Paths. Revolution in the School: A Project for Lifelong Learning. *International Journal of Digital Literacy and Digital Competence*, *7*(1), 1–18. doi:10.4018/IJDLDC.2016010101

Lin, L., & Lipsmeyer, B. (2015). The Environmental and Technological Factors of Multitasking. In A. Mesquita & C. Tsai (Eds.), *Human Behavior, Psychology, and Social Interaction in the Digital Era* (pp. 1–20). Hershey, PA: IGI Global. doi:10.4018/978-1-4666-8450-8.ch001

Lin, T., Wu, Z., & Chen, Y. (2015). Using High-Frequency Interaction Events to Automatically Classify Cognitive Load. In A. Mesquita & C. Tsai (Eds.), *Human Behavior, Psychology, and Social Interaction in the Digital Era* (pp. 210–228). Hershey, PA: IGI Global. doi:10.4018/978-1-4666-8450-8.ch010

Linton, J., & Stegall, D. (2015). Common Core Standards for Mathematical Practice and TPACK: An Integrated Approach to Instruction. In Standards and Standardization: Concepts, Methodologies, Tools, and Applications (pp. 92-107). Hershey, PA: IGI Global. doi:10.4018/978-1-4666-8111-8.ch005

Liu, J. C., & Kaye, E. R. (2016). Preparing Online Learning Readiness with Learner-Content Interaction: Design for Scaffolding Self-Regulated Learning. In L. Kyei-Blankson, J. Blankson, E. Ntuli, & C. Agyeman (Eds.), *Handbook of Research on Strategic Management of Interaction* (pp. 216–243). Hershey, PA: IGI Global. doi:10.4018/978-1-4666-9582-5.ch009

Liu, X., & Zhu, T. (2017). Comparing Online Personality of Americans and Chinese. In I. Management Association (Ed.), Gaming and Technology Addiction: Breakthroughs in Research and Practice (pp. 339-351). Hershey, PA: IGI Global. doi:10.4018/978-1-5225-0778-9.ch016

Lucas, K., & Murdock, J. (2014). Developing an Online Counseling Skills Course. *International Journal of Online Pedagogy and Course Design*, *4*(2), 46–63. doi:10.4018/ijopcd.2014040104

Lyublinskaya, I. (2015). Evolution of a Course for Special Education Teachers on Integrating Technology into Math and Science. In M. Niess & H. Gillow-Wiles (Eds.), *Handbook of Research on Teacher Education in the Digital Age* (pp. 521–547). Hershey, PA: IGI Global. doi:10.4018/978-1-4666-8403-4.ch020

Mac Mahon, B., Grádaigh, S. Ó., & Ghuidhir, S. N. (2016). iTE: Student Teachers using iPad on a Second Level Initial Teacher Education Programme. *International Journal of Mobile and Blended Learning*, *8*(2), 21–34. doi:10.4018/IJMBL.2016040102

Maina, E. M., Wagacha, P. W., & Oboko, R. (2016). Enhancing Active Learning Pedagogy through Online Collaborative Learning. In J. Keengwe & G. Onchwari (Eds.), *Handbook of Research on Active Learning and the Flipped Classroom Model in the Digital Age* (pp. 232–254). Hershey, PA: IGI Global. doi:10.4018/978-1-4666-9680-8.ch012

Marshall, A. M., & Callahan, K. M. (2016). Mathematics Teacher Educators' TPACK and MKT Knowledge Domains: Designing Online Discussion Blogs. In M. Niess, S. Driskell, & K. Hollebrands (Eds.), *Handbook of Research on Transforming Mathematics Teacher Education in the Digital Age* (pp. 353–380). Hershey, PA: IGI Global. doi:10.4018/978-1-5225-0120-6.ch014

Matthews, A., & Kitchin, R. M. Jr. (2015). Examining Design Pattern Strategies as a Means to Achieve Social Presence in the Online Classroom. In R. Wright (Ed.), *Student-Teacher Interaction in Online Learning Environments* (pp. 22–49). Hershey, PA: IGI Global. doi:10.4018/978-1-4666-6461-6.ch002

McBroom, E. S., Jiang, Z., Sorto, M. A., White, A., & Dickey, E. (2016). Dynamic Approach to Teaching Geometry: A Study of Teachers' TPACK Development. In M. Niess, S. Driskell, & K. Hollebrands (Eds.), *Handbook of Research on Transforming Mathematics Teacher Education in the Digital Age* (pp. 519–550). Hershey, PA: IGI Global. doi:10.4018/978-1-5225-0120-6.ch020

McGee, P. (2014). Blended Course Design: Wheres the Pedagogy? *International Journal of Mobile and Blended Learning*, *6*(1), 33–55. doi:10.4018/ijmbl.2014010103

Medárová, V. (2014). E-Learning Tools Applied in Teaching English at the University Level. In T. Issa, P. Isaias, & P. Kommers (Eds.), *Multicultural Awareness and Technology in Higher Education: Global Perspectives* (pp. 156–175). Hershey, PA: IGI Global. doi:10.4018/978-1-4666-5876-9.ch008

Mellati, M., & Khademi, M. (2015). The Impacts of Distance Interactivity on Learners Achievements in Online Mobile Language Learning: Social Software and Participatory Learning. *International Journal of Web-Based Learning and Teaching Technologies*, *10*(3), 19–35. doi:10.4018/ijwltt.2015070102

Mishra, P., Henriksen, D., & Mehta, R. (2015). Creativity, Digitality, and Teacher Professional Development: Unifying Theory, Research, and Practice. In M. Niess & H. Gillow-Wiles (Eds.), *Handbook of Research on Teacher Education in the Digital Age* (pp. 691–722). Hershey, PA: IGI Global. doi:10.4018/978-1-4666-8403-4.ch026

Modell, M. G. (2017). Learning to Lead Collaborative Student Groups to Success. In C. Martin & D. Polly (Eds.), *Handbook of Research on Teacher Education and Professional Development* (pp. 187–209). Hershey, PA: IGI Global. doi:10.4018/978-1-5225-1067-3.ch010

Mulder, D. J. (2016). Pre-Service Teachers and Technology Integration: International Cases and Generational Attitudes toward Technology in Education. In J. Keengwe, J. Mbae, & G. Onchwari (Eds.), *Handbook of Research on Global Issues in Next-Generation Teacher Education* (pp. 83–103). Hershey, PA: IGI Global. doi:10.4018/978-1-4666-9948-9.ch005

About the Authors

Justin Chase is an educational researcher, with primary research interests in motivation, and statistics education. He graduated from the University of Maine in 2009 with a B.A. in psychology, and an M.S. in psychological science in 2012 from Montana State University, and a certificate of advanced study in educational research from SUNY in 2016.

Zheng Yan is Associate Professor of Developmental Psychology at the University at Albany. His major publications include the *Encyclopedia of Cyber Behavior*, the *Encyclopedia of Mobile Phone Behavior*, and *Mobile Phone Behavior*. He has been a co-editor of the *International Journal of Cyber Behavior, Psychology and Learning* since 2012.

Index

www.ingramcontent.com/pod-product-compliance
Lightning Source LLC
LaVergne TN
LVHW062120200926
848484LV00007B/1725